MANAGING and MINING MULTIMEDIA DATABASES

MANAGING and MINING MULTIMEDIA DATABASES

Bhavani Thuraisingham

CRC Press

Boca Raton   London   New York   Washington, D.C.

Library of Congress Cataloging-in-Publication Data

Thuraisingham, Bhavani M.
    Managing and mining multimedia databases / Bhavani Thuraisingham.
        p. cm.
    Includes bibliographical references and index.
    ISBN 0-8493-0037-1
    1. Database management. 2. Data mining. 3. Multimedia systems. I. Title.

QA76.9.D3 T458 2001
006.7—dc21
                                                                    2001025368

Visit the CRC Press Web site at www.crcpress.com

Preface

Recent developments in information systems technologies have resulted in computerizing many applications in various business areas. Data has become a critical resource in many organizations; therefore efficient access to data, sharing or extracting information from the data, and making use of this information have become urgent needs. As a result, there have been many efforts to integrate the various data sources scattered across several sites and to extract information from these databases in the form of patterns and trends. These data sources may be databases managed by database management systems, or they could be warehoused in a repository from multiple sources. The advent of the World Wide Web (WWW) in the mid 1990s has resulted in even greater demand for managing data, information, and knowledge effectively. There is now so much data on the Web that managing it with conventional tools is becoming almost impossible. New tools and techniques are needed to effectively manage this data. Therefore, various tools are being developed to provide the interoperability and warehousing between the multiple data sources and systems, as well as to extract information from the databases and warehouses on the Web.

Data in Web databases are both structured and unstructured. Structured databases include relational and object databases. Unstructured databases include text, image, audio, and video databases. In general, multimedia databases are unstructured. Some text databases are semistructured, meaning that they have partial structure. Developments in multimedia database management systems have exploded during the past decade. While numerous papers and some texts have appeared in multimedia databases, more recently these databases are being mined to extract useful information. Furthermore, multimedia databases are being accessed on the Web. There is currently little information about providing a complete set of services for multimedia databases. These services include managing, mining, and integrating multimedia databases on the Web for electronic enterprises.

The focus of this book is on managing and mining multimedia databases for the electronic enterprise. We focus on database management system techniques for text, image, audio, and video databases. We then address issues and challenges regarding mining the multimedia databases to extract information that was previously unknown. Finally, we discuss the directions and challenges of integrating multimedia databases for the Web. In particular, e-business and its relationship to managing and mining multimedia databases will be discussed. Few texts provide a comprehensive set of services for multimedia data management, although numerous research papers have been published on this topic. The purpose of this book is to discuss complex ideas in multimedia data management and mining in a way that can be understood by someone who wants background information in this area. Technical managers as well as those interested in technology will benefit from this book. We employ a data-centric approach to describe multimedia technologies. The concepts are explained using e-commerce and the Web as an application area.

This book is divided into three parts. Part I describes multimedia database management. Without the underlying concepts such as querying and storage management, one cannot develop multimedia information management for the Web. We start with an overview of multimedia database system architectures and data models. This is followed by a discussion of some critical functions for multimedia database management. These functions include query processing, metadata management, storage management, and distribution.

Part II describes multimedia data mining. We discuss text, image, video, and audio mining. These discussions also provide overviews of text/information retrieval, image processing, video information retrieval, and audio/speech processing.

Part III describes multimedia on the Web. We start with a discussion of how multimedia databases may be integrated on the Web and then address multimedia data management and mining for e-business. We discuss some of the emerging technologies to support multimedia data management, e.g., collaboration, knowledge management, and training. Next, we discuss security and privacy issues for multimedia databases with the Web in mind. Finally, emerging standards as well as prototypes and products for multimedia data management and mining are explored.

Since a lot of background information is needed to understand the concepts in this book, six appendices are included. Appendix A provides an overview and framework for data management, showing where multimedia data management fits into this framework. We then provide a discussion of database systems technologies followed by a discussion of data mining technologies. These are discussed next. These include object-programming languages, object databases, object-based design and analysis, distributed objects, and components and framework, which all have applications in multimedia data management. Next, we discuss security issues, and finally, we provide an overview of Web technologies and e-commerce. Since multimedia on the Web will be a critical part of our lives and the Web is central to this book, we have also provided an introduction to the Web.

Although our first three books, *Data Management Systems: Evolution and Interoperation*; *Data Mining: Technologies, Techniques, Tools, and Trends*; and *Web Data Management and Electronic Commerce*, would serve as excellent sources of reference, this book is fairly self-contained. We have provided a reasonably comprehensive overview of the various background material necessary to understand multimedia databases in the six appendices. However, some of the details of this background information, especially on data management and mining, can be found in our previous texts.

We have tried to obtain current information on products and standards. However, as emphasized repeatedly in our books, vendors and researchers are continually updating their systems, and therefore information valid today may not be accurate tomorrow. We urge the reader to contact the vendors and get up-to-date information. Note that many of the products are trademarks of various corporations. If we know or have heard of such trademarks, we use capital italic letters for the product when it is first introduced. Again, due to the rapidly changing nature of the computer industry, we encourage the reader to contact the vendors to obtain up-to-date information on trademarks and ownership of the various products.

We have tried our best to obtain references from books, journals, magazines, and conference and workshop proceedings, and have given only a few Web page URLs as references. Although we tried to limit URLs as references, we found that it was almost impossible to write a current text without referencing them. Although URLs often contain excellent reference material, some may no longer be available even by the time this book is published. Therefore, we also encourage the reader to check the Web periodically for current information on multimedia data management developments, prototypes, and products. There are several conference series devoted to this topic.

We repeatedly use the terms data, data management, database systems, and database management systems here. We elaborate on these terms in one of the appendices. Data management systems are defined as systems that manage data, extract meaningful information, and make use of the information extracted. Therefore, data management systems include database systems, data warehouses, and data mining systems. Data could be structured, such as that found in relational databases, or unstructured, such as text, voice, imagery, and video. Numerous discussions in the past have attempted to distinguish between data, information, and knowledge. In our previous books on data management and mining, we did not attempt to clarify these terms. We simply stated that data could be just bits and bytes or it could convey some meaningful information to the user. However, considering the Web as well as increasing interest in data, information, and knowledge management as separate areas, this book takes a different approach by differentiating between these terms as much as possible. For our purposes, data usually represents some value like numbers, integers, or strings. Information is obtained when some meaning is associated with the data; for example, John's salary is $20,000. Knowledge is something acquired through reading and learning. That is, data and information can be transferred into knowledge when uncertainty about it is removed. Note that it is rather difficult to give exact definitions of data, information, and knowledge. Sometimes we will use these terms interchangeably. Our framework for data management helps clarify some of the differences. To be consistent with the terminology in our previous books, we will also distinguish between database systems and database management systems. A database management system is the component that manages a database containing persistent data. A database system consists of both the database and the database management system.

This book provides a fairly comprehensive overview of multimedia data management and mining technologies as well as their application to e-commerce/business applications. The book is written for technical managers and executives as well as for technologists interested in learning about the subject. The complicated ideas surrounding this topic are expressed in a simplified manner but still provide much information. Note that like many areas in data management, unless someone has practical experience carrying out experiments and working with the various tools, it is difficult to appreciate what tools exist and how to develop multimedia applications. Therefore, we encourage the reader to not only read the information in this book and take advantage of the references provided, but we also urge anyone who is interested in developing multimedia applications to work with existing tools.

Multimedia data management is still a relatively new technology and incorporates many other technologies. Therefore, as the various technologies integrate and mature, we can expect progress in this area. That is, not only can we expect tools and techniques to manage and mine multimedia databases, we can also expect tools for multimedia warehouses and multimedia repositories on the Web. We can look forward to rapid developments with respect to many of the ideas, concepts, and techniques discussed in this book. We urge the reader to stay current with all the developments in this emerging and useful technology area. This book is intended to provide background information as well as some of the key points and trends in multimedia data management on the Web.

It should be noted that e-commerce is one of the fastest growing technologies. Not only is there tremendous interest in text-based e-commerce, but we expect voice-based e-commerce to explode over the next few years. Furthermore, the models for e-commerce will also change due to the various laws and regulations that will develop. E-commerce will occur across states and countries, and therefore, state, federal, and international rules and regulations will have to be enforced.

There is so much to write about multimedia data management and the Web that we could have written this book forever. While we have tried to provide as much information as possible, there is so much more to write about. We hear about e-commerce daily on the news, various television programs, and in conversation, and the amount of information on this topic can only increase as we enter the new millennium. We advise the reader to keep up with developments, determine what is important and what is not, and be knowledgeable about this subject. It will be helpful not only in our business lives and careers, but also in our personal lives in terms of investments, travel, selecting schools, and many other activities.

The views and conclusions expressed in this book are those of the author and do not reflect the views, policies, or procedures of the author's institution or sponsors.

Author

Bhavani Thuraisingham, Ph.D., recipient of the IEEE Computer Society's prestigious 1997 Technical Achievement Award for her outstanding and innovative work in secure data management, is a chief scientist in data management at MITRE Corporation's Information Technology Directorate in Bedford, Massachusetts. In this capacity, she provides technology directions in data, information, and knowledge management for the Information Technology Directorate of MITRE's Air Force Center. In addition, she is also an expert consultant in computer software to MITRE's work for the Internal Revenue Service. Her current work focuses on data mining as it relates to multimedia databases and database security, distributed object management with emphasis on real-time data management, and Web data management applications in electronic commerce. She also serves as adjunct professor of computer science at Boston University and teaches a course in advanced data management and data mining.

Prior to beginning her current position at MITRE in May 1999, she was the department head in data management and object technology in MITRE's Information Technology Division in the Intelligence Center for four years. In that position, she was responsible for the management of about 30 technical staff in four key areas: distributed databases, multimedia data management, data mining and knowledge management, and distributed objects and quality of service. Prior to that, she held various technical positions including lead, principal, and senior principal engineer, and was head of MITRE's research in evolvable interoperable information systems as well as data management and co-director of MITRE's Database Specialty Group. She managed fifteen research projects under the Massive Digital Data Systems effort for the intelligence community and was also a team member of the AWACS modernization research project from 1993 to 1999. Before that, she led team efforts on the designs and prototypes of various secure database systems for government sponsors between 1989 and 1993.

Prior to joining MITRE in January 1989, Dr. Thuraisingham worked in the computer industry from 1983 to 1989. She was first a senior programmer/analyst with Control Data Corporation for over two years, working on the design and development of the CDCNET product, and later she was a principal research scientist with Honeywell Inc. for over three years, conducting research, development, and technology transfer activities. She was also an adjunct professor of computer science and a member of the graduate faculty at the University of Minnesota between 1984 and 1988. Prior to starting her industrial experience and after completing her Ph.D., she was a visiting faculty member, first in the Department of Computer Science, at the New Mexico Institute of Technology, and then at the Department of Mathematics at the University of Minnesota between 1980 and 1983. Dr. Thuraisingham earned a B.Sc., M.Sc., M.S. and also received her Ph.D. degree from the United Kingdom at the age of 24. She is a senior member of the IEEE and a member of the ACM, British Computer Society, and AFCEA. She has a certification in Java programming and has also completed a management development program.

Dr. Thuraisingham has published over 350 technical papers and reports, including over 50 journal articles, and is the inventor of three U.S. patents for MITRE on database inference control. She also serves on the editorial boards of various journals, including *IEEE Transactions on Knowledge and Data Engineering*, the *Journal of Computer Security*, and *Computer Standards and Interfaces Journal*. She gives tutorials in data management, including data mining, object databases, and Web databases, and currently teaches courses at both the MITRE Institute and the AFCEA Educational Foundation. She has chaired or co-chaired several conferences and workshops including IFIP's 1992 Database Security Conference, ACM's 1993 Object Security Workshop, ACM's 1994 Objects in Healthcare Information Systems Workshop, IEEE's 1995 Multimedia Database Systems Workshop, IEEE's 1996 Metadata Conference, AFCEA's 1997 Federal Data Mining Symposium, IEEE's 1998 COMPSAC Conference, IEEE's 1999 WORDS Workshop, IFIP's 2000 Database Security Conference, and IEEE's 2001 ISADS Conference. She is a member of OMG's real-time special interest group, founded the C4I special interest group, and has served on panels in the field of data management and mining. She has edited several books as well as special journal issues and was the consulting editor of the Data Management Handbook series by CRC's Auerbach Publications in 1996 and 1997. She is the author of the books *Data Management Systems Evolution and Interoperation*; *Data Mining: Technologies, Techniques, Tools and Trends*; and *Web Data Management and Electronic Commerce*, published by CRC Press.

Dr. Thuraisingham has given invited presentations at several conferences including recent keynote addresses at the Second Pacific Asia Data Mining Conference 1998, SAS Institute's Data Mining Technology Conference 1999, IEEE Artificial Neural Networks Conference 1999, and IEEE Tools in AI Conference 1999. She has also delivered the featured addresses at AFCEA's Federal Database Colloquium from 1994 through 2000. She has given presentations worldwide, including in the United States, Canada, United Kingdom, France, Germany, Italy, Spain, Switzerland, Austria, Belgium, Sweden, Finland, Denmark, Norway, The Netherlands, Greece, Ireland, Egypt, South Africa, India, Hong Kong, Taiwan, Japan, Singapore, New Zealand, and Australia. She also gives seminars and lectures at various universities around the world including the University of Cambridge in England and the Massachusetts Institute of Technology, and participates in panels at the National Academy of Sciences, and the Air Force Scientific Advisory Board.

Acknowledgment

I thank my management for providing an environment where it is exciting and challenging to work, my professors and teachers for having given me the foundations upon which to build my skills, my sponsors and colleagues, all others who have supported my education and my work, and especially those who have reviewed various portions of this book. Last but not least, I thank the two most important people in my life: my husband Thevendra and my son Breman for giving me so much encouragement to write this.

Bhavani Thuraisingham, Ph.D.
Bedford, Massachusetts

Dedication

To my dear friend
Martha Lewandowski

Table of Contents

1 Introduction

1.1 TRENDS

Recent developments in information systems technologies have resulted in computerizing many applications in various business areas. Data has become a critical resource in many organizations, therefore efficient access to data, sharing or extracting information from the data, and making use of this information have become urgent needs. As a result, many efforts to integrate the various data sources scattered across several sites as well as extract information from these databases in the form of patterns and trends. These data sources may be databases managed by database management systems, or they could be data warehoused in a repository from multiple data sources. The advent of the World Wide Web (WWW) in the mid 1990s has resulted in even greater demand for managing data, information, and knowledge effectively. There is now so much data on the Web that managing it with conventional tools is becoming almost impossible. New tools and techniques are needed to effectively manage these data. Therefore, various tools are being developed to provide interoperability and warehousing between multiple data sources and systems, as well as to extract information from the databases and warehouses on the Web.

Data in Web databases are both structured and unstructured. Structured databases include those that have some structure such as relational and object databases. Unstructured databases include those that have very little structure such as text, image, audio, and video databases. In general, multimedia databases are unstructured. Some text databases are semistructured databases, meaning that they have partial structure. The developments in multimedia database management systems have exploded during the past decade. While numerous papers and some texts have appeared in multimedia databases, more recently these databases are being mined to extract useful information. Furthermore, multimedia databases are being accessed on the Web. That is, there is currently little information about providing a complete set of services for multimedia databases. These services include managing, mining, and integrating multimedia databases on the Web for an electronic enterprise.

The focus of this book is on managing and mining multimedia databases for the electronic enterprise. We focus on database management system techniques for text, image, audio, and video databases. We then address issues and challenges regarding mining the multimedia databases to extract information that was previously unknown. Finally, we discuss the directions and challenges of integrating multimedia databases for the Web. In particular, e-business and its relationship to managing and mining multimedia databases will be discussed. As mentioned earlier, there are hardly any texts on providing a comprehensive set of services for multimedia data management, although numerous research papers have been published on this topic. The purpose of this book is to discuss complex ideas in multimedia data management and mining in a way that can be understood by someone who wants background

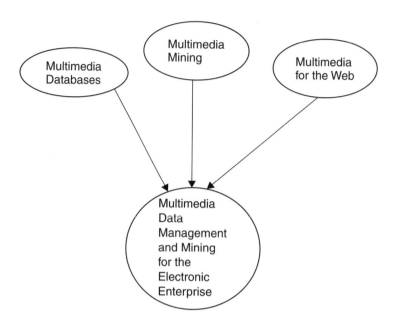

FIGURE 1.1 Multimedia data management and mining for the electronic enterprise.

information in this area. Technical managers as well as those interested in technology will benefit from this book. We employ a data-centric approach to describe multimedia technologies. The concepts are explained using e-commerce and the Web as an application area.

The organization of this chapter is as follows. Multimedia data management issues are discussed in Section 1.2. Multimedia data mining is the subject of Section 1.3. Applications of multimedia data management and mining for an electronic enterprise are discussed in Section 1.4. In particular, multimedia on the Web and multimedia for e-business are discussed. Note that Sections 1.2, 1.3, and 1.4 are elaborated in Parts I, II, and III of this book, as illustrated in Figure 1.1. The organization of this book is the subject of Section 1.5. To put this all together, a framework for multimedia data management and mining is described which helps us give some context to the various Web data management technologies. Finally, the chapter is summarized in Section 1.6, which also includes a discussion of directions.

1.2   MULTIMEDIA DATABASE MANAGEMENT

A multimedia database system is comprised of a multimedia database management system (MM-DBMS) that manages a multimedia database, which is a database containing multimedia data. Multimedia data may include structured data as well as semistructured and unstructured data such as voice, video, text, and images. That is, an MM-DBMS provides support for storing, manipulating, and retrieving multimedia data from a multimedia database. In a certain sense, a multimedia database system is a type of heterogeneous database system because it manages heterogeneous data types.

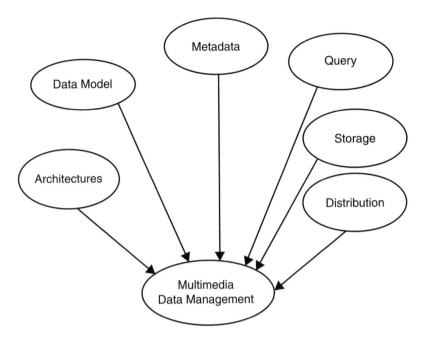

FIGURE 1.2 Multimedia database management.

An MM-DBMS must provide support for typical database management system functions. These include query processing, update processing, transaction management, storage management, metadata management, security, and integrity. In addition, in many cases, the various types of data such as voice and video have to be synchronized for display, and, therefore, real-time processing is also a major issue in an MM-DBMS. Figure 1.2 illustrates some of the key functions that will be addressed in Part II.

MM-DBMSs are becoming popular for various applications including C4I, CAD/CAM, air traffic control, and, particularly, entertainment. While the terms multimedia and hypermedia are often used interchangeably, we differentiate between the two. While an MM-DBMS manages a multimedia database, a hypermedia DBMS not only manages a multimedia database, but also provides support for browsing the database by following links. That is, a hypermedia DBMS contains an MM-DBMS.

Recently, there has been much research on designing and developing MM-DBMSs, and, as a result, prototypes and some commercial products are now available.[3,4,21,61-63,92,102,129] However, as stated by Dao and Thuraisingham,[124] there are several areas that need further work. Research on developing an appropriate data model to support data types such as video is needed. Some experts have proposed object-oriented database management systems (OO-DBMS) for storing and managing multimedia data because they have been found to be more suitable for handling large objects and multimedia data such as sound and video which consume considerable storage space.[139] Although such systems show some promise, they are not sufficient to capture all of the requirements of multimedia applications. For example, in many cases, voice and video data which may be stored in objects have to be synchronized when displayed. The constraints for synchronization are not specified in the object models. Another area that needs research is the development of efficient techniques for indexing. Data manipulation operations such as video editing are still in the early stages. Furthermore, the multimedia databases need to be integrated for many applications as they are distributed. For example, audio data in database 1 has to be integrated with video data in database 2 and displayed to the analyst. Various aspects of such integration will be covered in Part I of this book.

1.3 MULTIMEDIA DATA MINING

Recently, there has been much interest in mining multimedia databases such as text, images, and video. As mentioned, many data mining tools work on relational databases. However, a considerable amount of data is now in multimedia format. There is a large amount of text and image data on the Web. News services provide a lot of video and audio data. This data has to be mined so that useful information can be extracted. One solution is to extract structured data from the multimedia databases and then mine the structured data using traditional data mining tools. Another solution is to develop mining tools to operate on the multimedia data directly. Note that to mine multimedia data, we must mine combinations of two or more data types, such as text and video, or text, video, and audio. However, in this book we deal mainly with one data type at a time because we first need techniques to mine the data belonging to the individual data types before mining multimedia data. In the future, tools for multimedia data mining will probably be developed.

As stated earlier, multimedia data includes text, images, video, and audio. Text and images are still media, while audio and video are continuous media. The issues surrounding still and continuous media are somewhat different and will be explained in Part I of this book. Part II will look at text, image, video, and audio and consider how such data can be mined. First of all, what are the differences between mining multimedia data and topics such as text, image, and video retrieval? What is meant by mining such data? What are the developments and challenges? Note that Part II elaborates on each of these topics. Figure 1.3 illustrates multimedia data mining, in particular, various aspects of multimedia data mining.

Data mining has an impact on the functions of multimedia database systems. For example, the query processing strategies have to be adapted to handle mining queries if there is a tight integration between the data miner and the database system. This will then have an impact on the storage strategies. Furthermore, the data model will also have an impact. At present, many of the mining tools work on relational databases. However, if object-relational databases are to be used for multimedia modeling, then data mining tools have to be developed to handle such databases.

1.4 MULTIMEDIA FOR THE WEB AND
THE ELECTRONIC ENTERPRISE

There are various supporting technologies for the Web. Many of them are discussed by Thuraisingham.[128] One of the key supporting technologies is database systems. There is a tremendous amount of data on the Web; some of it stored in files and

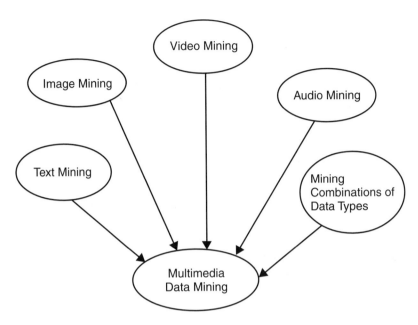

FIGURE 1.3 Multimedia data mining.

some in databases. This data has to be managed effectively. Therefore, query processing, transaction management, storage management, and metadata management all play key roles in Web data management.

Another technology that is becoming critical for the Web is data mining. Data mining is the process of forming conclusions from premises often previously unknown from large quantities of data. There are two aspects: one is to mine data on the Web and extract useful information, and the other is to mine Web usage patterns to give guidance to the user.

Since multimedia has an impact on both databases and data mining, multimedia database management and data mining are key technologies for the Web. Using the Web involves handling different media types such as voice, text, and audio. In addition, the multimedia data has to be mined. Finally, multimedia data mining can help carry out targeted marketing for e-business operations.

Figure 1.4 illustrates multimedia data mining and management on the Web. More details are given in Part III of this book.

1.5 ORGANIZATION OF THIS BOOK

This book covers the essential topics in multimedia data management and data mining for an electronic enterprise in three parts: multimedia databases, multimedia data mining, and multimedia data management and mining for the electronic enterprise. Figure 1.5 illustrates a multimedia data management framework. This framework has three layers. Layer I is the multimedia data management layer. It describes the various multimedia database technologies such as architectures, models, query,

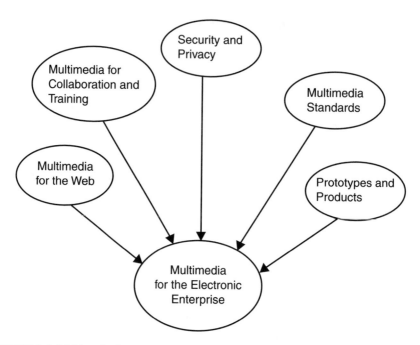

FIGURE 1.4 Multimedia for the electronic enterprise.

metadata, storage, and distribution. Layer II is the multimedia data mining layer. This layer describes the various data mining technologies including text mining, image mining, video mining, and audio mining. Layer III is the multimedia for electronic enterprise layer, and it describes multimedia technologies for the Web and e-business. In addition, standards, prototypes, and products are discussed.

The layers are described in three parts in this book. Part I, consisting of six chapters, describes the various multimedia data management technologies. Chapter 2 describes multimedia database architectures including loose coupling and tight coupling architectures. Chapter 3 describes data modeling for multimedia databases including relational and object models. Chapter 4 discusses metadata issues including a definition of metadata for multimedia databases and then focuses on how this metadata could be managed. Chapter 5 discusses query technologies for multimedia databases including query strategies and languages. Chapter 6 focuses on storage technologies for multimedia data and provides a discussion of access methods and indexing. Finally, Chapter 7 describes distribution issues for multimedia databases.

Part II, consisting of two chapters, addresses multimedia data mining. Chapter 8 provides a general overview of multimedia data mining. The technologies and techniques for data mining discussed by Thuraisingham[127] are examined, as is the impact of mining multimedia data on these technologies and techniques. Chapter 9 discusses the issues on mining text, images, video, and audio data. Note that in general, multimedia data mining means mining combinations of data types. However, we need to get a good handle on mining the individual data types first before we

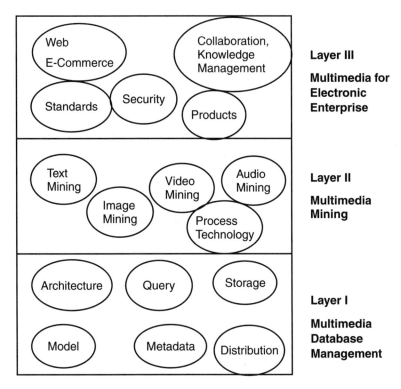

FIGURE 1.5 Framework for multimedia data management and mining for the electronic enterprise.

can handle combinations of data types. Therefore, Chapter 9 focuses mainly on mining individual data types.

While Parts I and II address multimedia data management and mining technologies, Part III addresses the important application area for multimedia data management and mining — the electronic enterprise. An electronic enterprise is an enterprise that uses state-of-the-art technologies surrounding the Web and carries out activities such as e-business and e-commerce. Part III consists of Chapters 10 through 13. Chapter 10 provides an overview of multimedia for the Web and then shows how multimedia technologies may support e-business. Chapter 11 discusses how multimedia can support applications such as collaboration, knowledge management, training, and entertainment. Chapter 12 addresses security and privacy aspects for data mining and multimedia data. Chapter 13 provides an overview of some of the emerging multimedia standards for the enterprise as well as an overview of various multimedia prototypes and products.

Chapter 14 summarizes the book and provides a discussion of challenges and directions. Each of the chapters in Parts I, II, and III (Chapters 2 through 13) starts with an overview and ends with a summary. Each part also begins with an introduction and ends with a conclusion. Finally, the book includes six appendices that provide

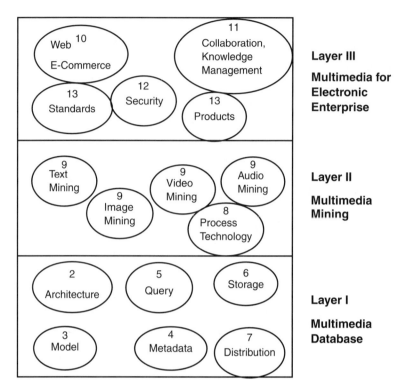

FIGURE 1.6 Components addressed in this book. Numbers shown are chapter numbers in which the relevant information can be found.

useful background information. As the reader will see, both data management and data mining technologies play a major role in multimedia data management and mining. Appendix A provides an overview of trends in data management technology. Appendix B provides an overview of the developments and trends in database systems as well as in distributed database systems. Data mining is discussed in Appendix C. Object technology is the subject of Appendix D. Security issues are discussed in Appendix E. An introduction to the Web and e-business is given in Appendix F. Figure 1.6 illustrates all the components addressed in this book.

1.6 HOW DO WE PROCEED?

This chapter has provided an introduction to multimedia data management and mining for the electronic enterprise. We first discussed multimedia database management architectures, models, and functions, which include query processing, metadata management, and storage management. We then discussed multimedia data mining including text, video, image, and audio mining. Finally, this chapter showed how multimedia technologies support an electronic enterprise including e-business and e-commerce enterprises. Parts I, II, and III of this book elaborate on Sections 1.2, 1.3, and 1.4, respectively. The book's organization is detailed in Section 1.5, which also includes a framework for organization purposes. Our framework is a three-layer framework, and each layer is addressed in a different part of this book.

This book provides the information for a reader to get familiar with multimedia data management and data mining. Many important topics are covered so that the reader has some idea as to what multimedia data management and mining is all about. For an in-depth understanding of the various topics covered in this book, we recommend the reader to the references provided. Various papers and articles have appeared on multimedia data management and related areas. Many of these are referenced throughout this book. Some interesting discussions have been published in the proceedings of the IEEE Multimedia Database Workshop Series in 1995, 1996, and 1998.[61-63]

There is so much to write about multimedia data management and its application to e-commerce that we could continue writing this book forever. That is, while we have tried to provide as much information as possible in this book, there is so much more to write about. We hear about e-commerce and multimedia e-commerce daily on the news, various television programs, and in conversation, and the amount of information on this topic can only increase as we enter the new millennium. Readers should keep up with the developments, discern what is important and what is not, and be knowledgeable about this subject. It will be helpful not only in our business lives, but also in our personal lives, for example, personal investments and other activities.

Part I

Managing Multimedia Databases

INTRODUCTION

Part I, consisting of six chapters, describes multimedia database system technologies. Chapter 2 provides an overview of architectures for multimedia database management. Different dimensions are discussed. We first describe loose coupling versus tight coupling architectures and then take a look at schemas architectures. Functional architectures are discussed next. Finally, we address distribution and interoperability architectures.

Chapter 3 provides an overview of data modeling for multimedia databases. We examine both object and relational models and also explore some other models. A good data model is essential for implementing efficient multimedia database systems.

Chapter 4 discusses metadata for multimedia databases. We first define different types of metadata. For example, metadata for image data may include descriptions of images as well as annotations of images. Metadata for video data may include snapshots of the video. We also discuss metadata management aspects for multimedia databases.

Chapter 5 describes querying multimedia databases. In particular, query processing and optimization issues as well as query language aspects will be discussed. Issues closely related to querying include transaction management and multimedia object editing, which are also addressed in this chapter.

Chapter 6 describes storage aspects for multimedia databases. These include storage mechanisms for text, images, video, and audio data as well as access methods and indexing strategies for efficient multimedia data retrieval.

Finally, Chapter 7 addresses distributed multimedia database systems. We examine distributed architectures and explore functions such as distributed query management.

While we discuss some of the important functions of multimedia databases in Part I, there are many other aspects such as security, integrity, and quality of service processing for multimedia databases. Some of these issues will be addressed in various parts of this book.

2 Architectures for Multimedia Database Systems

2.1 OVERVIEW

Various architectures are being examined to design and develop a multimedia database management system (MM-DBMS). These architectures fall under different categories, and this chapter examines the various types of architectures.

One architecture type involves integrating multimedia data with the database system. There are two approaches. In the loose coupling approach, the multimedia data is managed by the file system, while the database system manages the metadata. In the tight coupling approach, the multimedia data is managed by the database system. Another type of architecture is schema architecture. For example, does the three-schema architecture apply for a multimedia database system? A third type of architecture is functional architecture, describing the functions of a multimedia database system. A fourth type of architecture is whether a multimedia database system extends a traditional database system. This is what we call a system architecture. A fifth type of architecture is a distributed architecture, where a multimedia database is distributed. Finally, multimedia databases may be heterogeneous in nature and need to be integrated. The architecture for integrating heterogeneous databases is known as interoperable architecture. Figure 2.1 illustrates the various types of architectures.

Section 2.2 describes loose coupling versus tight coupling architecture. Schema architecture is discussed in Section 2.3, and functional architecture in Section 2.4. System architecture is discussed in Section 2.5, and distributed architectures in Section 2.6. Section 2.7 describes interoperable architecture, and finally, in Section 2.8, we discuss architectures for hypermedia database systems. The chapter is summarized in Section 2.9.

2.2 LOOSE COUPLING VERSUS TIGHT COUPLING

This section describes the loose coupling versus tight coupling approaches to designing a multimedia database system. In the loose coupling approach, the DBMS is used to manage only the metadata, and a multimedia file manager is used to manage the multimedia data. Then there is a module for integrating the DBMS and the multimedia file manager. Figure 2.2 illustrates loose coupling architecture. In Figure 2.2, the MM-DBMS consists of three modules: the DBMS managing the

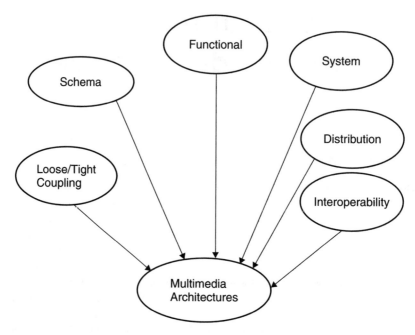

FIGURE 2.1 Types of architectures.

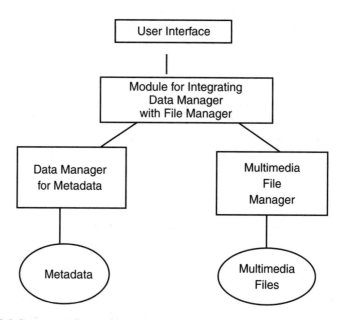

FIGURE 2.2 Loose coupling architecture.

metadata, the multimedia file manager, and the module for integrating the two. The advantage of the loose coupling approach is that one can use various multimedia file systems to manage the multimedia data.

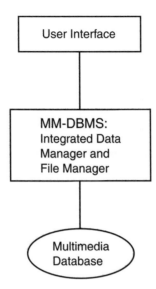

FIGURE 2.3 Tight coupling architecture.

The second architecture, illustrated in Figure 2.3, is the tight coupling approach. In tight coupling architecture, the DBMS manages both the multimedia database and the metadata. That is, the DBMS is an MM-DBMS. Tight coupling architecture is advantageous because all DBMS functions can be applied on the multimedia database. This includes query management, transaction processing, metadata management, storage management, and security and integrity management. Note that with the loose coupling approach, unless the file manager performs the DBMS functions, the DBMS only manages the metadata for the multimedia data.

Much of the discussion in this book assumes a tight coupling design. That is, the MM-DBMS manages the multimedia database and performs various functions such as query processing and storage management.

2.3 SCHEMA ARCHITECTURE

Schema architectures can be described in various ways with respect to different characteristics. Schema is essentially the metadata that describes the multimedia data. One can directly apply the three-schema architecture discussed in Appendix B for multimedia database systems. Here, the external schema will define the views that users have of the database, such as video or audio views. The logical schema is based on the data model for the multimedia database. This data model will be the subject of Chapter 3. Internal schema are the internal data structures, for example, variation of B trees for multimedia databases. An example of such an index structure is an R+ tree, and that concept will be discussed in Chapter 6. Three-schema architecture for multimedia databases is illustrated in Figure 2.4.

One can also look at schema from another point of view. Instead of multimedia data, assume that individual data types are stored in separate databases. For example,

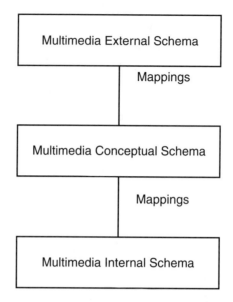

FIGURE 2.4 Three-schema architecture.

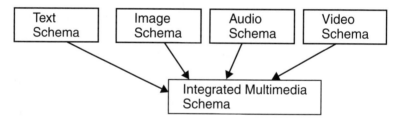

FIGURE 2.5 Integrated schema architecture.

video schema will describe the video database and audio schema will describe the audio database. Figure 2.5 illustrates the integration of the various types of schema.

2.4 FUNCTIONAL ARCHITECTURES

Figure 2.6 illustrates a functional architecture for an MM-DBMS. Functions of a multimedia database system include data representation, distribution, query/update processing, browsing and editing, quality of service processing, real-time scheduling, metadata management, storage management, and security/integrity management. Various aspects of these functions are discussed in subsequent chapters.

Figure 2.7 illustrates a more detailed view of functional architecture; the major modules are shown. The presentation layer presents various media types to the user, and the query manager performs query processing. The storage manager accesses the multimedia database, and the metadata manager manages the metadata. The interactions between the modules are also illustrated. Note that this is a slight

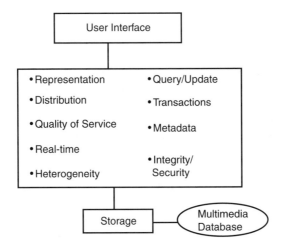

FIGURE 2.6 Functional architecture.

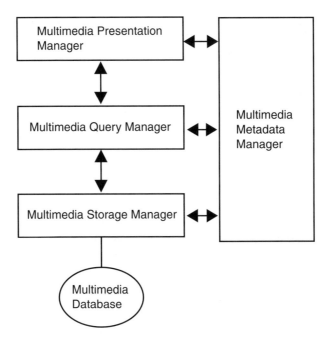

FIGURE 2.7 Modules of the functional architecture.

variation of the functional architecture discussed in Appendix B. Also note that there are other modules, such as transaction manager and security/integrity manager, that are not illustrated in Figure 2.7. For example we could include a transaction manager parallel to the query manager. Security and integrity managers will have to perform functions at all layers.

2.5 SYSTEM ARCHITECTURE

There are also other aspects of architectures. For example, a multimedia database system could use a commercial database system such as an object-oriented DBMS (OO-DBMS) to manage multimedia objects. However, relationships between objects and the representation of temporal relationships may involve extensions to the database management system. That is, a DBMS plus an extension layer provide complete support to manage multimedia data. In the alternative case, both the extensions and the database management functions are integrated so that there is one database management system to manage multimedia objects as well as the relationships between the objects. These two types of architectures are illustrated in Figures 2.8 and 2.9. We call them system architectures.

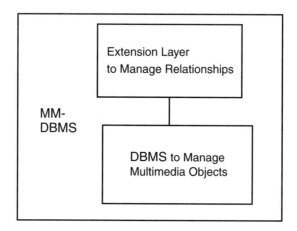

FIGURE 2.8 DBMS plus extension layer.

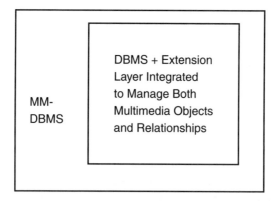

FIGURE 2.9 DBMS and extension layer integrated.

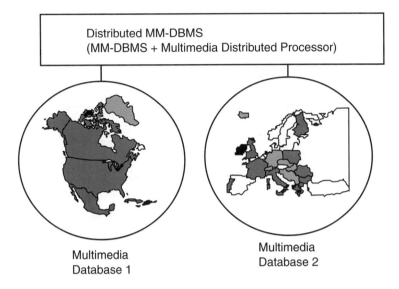

FIGURE 2.10 Data distribution.

2.6 DISTRIBUTED ARCHITECTURE

An MM-DBMS could also be distributed. An example of a distributed MM-DBMS is illustrated in Figure 2.10. In that example, one node has a map of the U.S., while another node has a map of Europe. Such a system must manage the distributed multimedia database. One can merge the two objects at the two nodes to obtain the result to a query such as "retrieve all the contents of the distributed database." The result will be an object which contains maps of both the U.S. and Europe.

Figure 2.11 illustrates another example of a distributed MM-DBMS. As shown in Figure 2.11, a distributed MM-DBMS is essentially a collection of MM-DBMSs connected through a network. The multimedia distributed processor module is responsible for handling data distribution issues as well as distributed query, transactions, metadata, and security and integrity management. For example, how are objects distributed? Are they distributed based on clusters or related topics? How can different objects be combined in query processing? Are there special mechanisms for distributed transaction management? How is the distributed metadata maintained? Recently, there have been some research efforts on distributed MM-DBMSs.[84] Many of these efforts assume a homogeneous environment except for different data types. Distributed MM-DBMSs could also be heterogeneous with respect to data models and other aspects. Distributed multimedia databases are discussed in Chapter 7.

2.7 INTEROPERABILITY ARCHITECTURE

Note that an MM-DBMS can also be based on a client-server approach. If this is the case, then a distributed object management (DOM) system such as the one based on common object request broker architecture (CORBA) can be used to integrate the various components. The special extensions for the CORBA interface definition

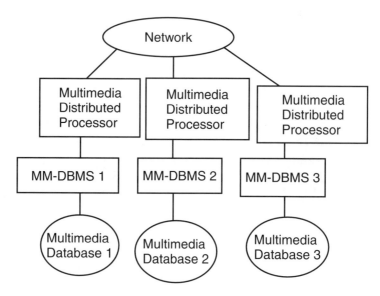

FIGURE 2.11 Distributed MM-DBMS.

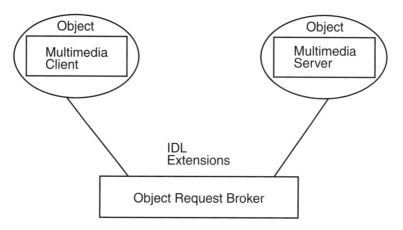

FIGURE 2.12 An architecture for interoperability.

language (IDL) to support multimedia data are being specified. A CORBA-based approach to designing an MM-DBMS is illustrated in Figure 2.12. MM-DBMSs can also form federations. For more information on heterogeneous database integration, refer to Thuraisingham.[126] An overview of distributed object management is provided in Appendix B.

Three-tier architectures are also popular for data and information management. In such an approach, the client tier has the presentation modules. The business objects comprise the middle tier, and the server tier carries out the database functions. Such an architecture is also becoming popular for multimedia applications. An example of three-tier architecture is illustrated in Figure 2.13.

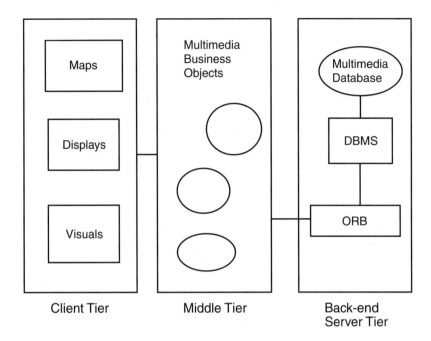

FIGURE 2.13 Three-tier architecture.

Finally, Geppert and Dittrich have described component-based architectures for database management.[45] They call such systems component database management systems (CDBMS). They argue that CDBMSs will be especially useful for complex applications such as multimedia applications. The idea is for database systems to be componentized so that each component can carry out different functions, as illustrated in Figure 2.14. Variations of CDBMSs have also been proposed by Geppert and Dittrich.[45]

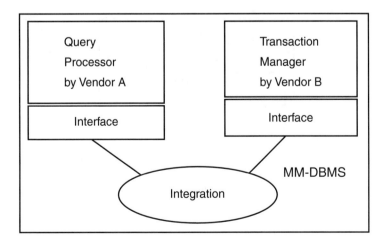

FIGURE 2.14 Component integration for multimedia database management.

The idea behind component integration is to put together a system from existing components. Some early work was carried out under the Defense Advanced Research Projects Agency (DARPA) funded initiative called Open OODB carried out by Texas Instruments.[112] The concept is that vendors take an all-or-nothing approach. That is, you get the entire system or none at all. What is needed, however, is a flexible system based on components. This means that the various components, such as query processor and transaction manager, can be developed by different vendors. A customer gets these components and puts them together. Therefore, the components have to meet well-defined interfaces. This approach is illustrated in Figure 2.14.

2.8 HYPERMEDIA ARCHITECTURE

The terms multimedia and hypermedia are often used interchangeably. This book, however, differentiates between the two. While an MM-DBMS manages a multimedia database, a hypermedia DBMS not only manages a multimedia database, but also provides support for browsing the database by following links. That is, a hypermedia DBMS includes an MM-DBMS. We have illustrated both an MM-DBMS and a hypermedia DBMS in Figures 2.15 and 2.16, respectively.

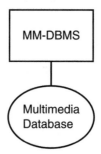

FIGURE 2.15 Multimedia database management system.

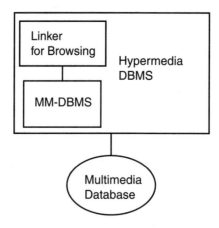

FIGURE 2.16 Hypermedia database management system.

As illustrated, a hypermedia DBMS includes an MM-DBMS and a linker. The linker module provides support for browsing the multimedia database by following links. This book focuses mainly on MM-DBMSs; hypermedia issues were discussed in our previous book, *Web Data Management and Electronic Commerce*, and will also be addressed in Part III of this book. Essentially, multimedia on the Web is mostly about hypermedia technology where multimedia database systems are augmented with browsers.

2.9 SUMMARY

This chapter discusses various types of architectures for multimedia database systems; in particular, loose/tight integration architecture, schema architecture, system architecture, functional architecture, distributed architecture, interoperability architecture, and hypermedia architecture. Essentially, we examined architectures for database systems and discussed the impact of multimedia data management on these architectures.

In the case of interoperability architecture, we examined various aspects including architectures for integrated heterogeneous databases, architectures based on object request brokers, client-server-based architectures, three-tier architectures, and component-based architectures. We believe that component-based architectures will gain popularity as more and more applications deal with complex data such as voice, text, video, and images.

There is no ideal architecture for a multimedia database system. The architecture selected will depend on the user needs and the application in question. Therefore, before building a multimedia database system, one needs to carry out an architecture study and then select the most appropriate architecture.*

---

* Background information for architectures is given in Appendix B. The reader is also referred to the following references: Bell and Grimson,[15] Ceri and Pelagatti,[20] Korth and Silberschatz,[78] and Oszu and Valdurez.[99]

3 Multimedia Data and Information Models

3.1 OVERVIEW

Various data models are being examined to design and develop MM-DBMSs, including relational, object, object-relational, and other semantic models. Relational models capture the relationships between multimedia objects, while object models capture the complex data structures. Object-relational models can capture both relationships and complex data structures. Semantic models and hypersemantic models can capture the rules that are enforced between the different media types; for example, "object A and object B have to play together."

In the early to mid-1990s, there was much debate as to which models would be appropriate for multimedia databases.[3] The competing models were object and relational models. After much discussion, experts now feel that one needs both objects as well as relationships to capture the complex data types and the relationships between them. In addition, rule-based models are useful to specify the timing constraints between objects. Therefore, a combination of objects, relations, and rules is needed. Figure 3.1 illustrates the various data models for multimedia databases.

This chapter examines various types of data models. Section 3.2 describes relational, object-relational, and semantic data models. Information modeling is the subject of Section 3.3. Information models can be regarded as data models combined with process models. That is, these models have to describe the data as well as the activities. The chapter is summarized in Section 3.4. Note that we have provided only the essential points of data modeling. A more comprehensive survey of spatio-temporal data models for multimedia databases is provided by Vazirgiannis and Sellis.[133]

3.2 DATA MODELING

3.2.1 OVERVIEW

In representing multimedia data, several features have to be supported. First of all, there has to be some way to capture the complex data types and all the relationships between the data. Various temporal constructs such as play-before, play-after, and play-together must be captured (see, for example, the discussion in Oomoto and Tanaka[94]). Figure 3.2 illustrates a representation of a multimedia database. In that representation, there are two objects, A and B. Object A consists of 2000 frames, and Object B consists of 3000 frames. Object A consists of a time interval between 4/95 and 8/95, and Object B consists of a time interval between 5/95 and 10/95.

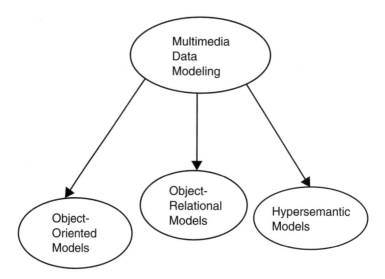

FIGURE 3.1 Data representation.

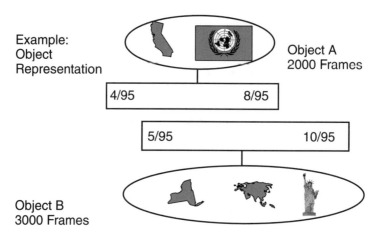

FIGURE 3.2 Data representation.

An appropriate data model is critical for representing a multimedia database. Relational, object-oriented, and object-relational data models have been examined to represent multimedia data. Some argue that relational models are better because they can capture relationships, while others argue that object models are better because they represent complex structures. Still others argue that hypersemantic data models are better because they capture both objects, relationships, and rules. We discuss all these models in the following subsections.

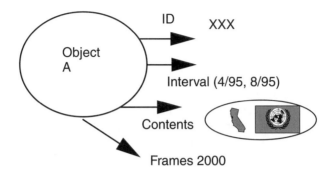

FIGURE 3.3 Data representation with object model.

ID	Interval	Contents	Frame
XXX	(4/95, 8/95)		2000

FIGURE 3.4 Data representation with object-relational model.

3.2.2 OBJECT VERSUS OBJECT-RELATIONAL DATA MODELS

In the example of Figure 3.2, with an object-oriented data model, each object in the figure corresponds to an object in the object model. The attributes of an object may be represented as instance variables and will include time interval, frames, and content description. In a relational model, the object corresponds to an instance of a relation. However, with atomic values, it is difficult to capture the attributes of the instance. In the case of an object-relational model, the attribute value of an instance may be an object. That is, for the instance that represents, for example, Object A, the attribute value time interval would be the pair (4/95, 8/95). Representing Object A with an object model is illustrated in Figure 3.3. Representing the same object with an object-relational model is illustrated in Figure 3.4.

Note that one could build extensions to an existing data model to support complex relationships for multimedia data. These relationships may include temporal relationships between objects such as play-together, play-before, and play-after. These relationships are discussed in the next subsection.

As mentioned earlier, some argue that object models are better for multimedia databases, while others argue that object-oriented models are better because they can represent complex data types. It appears that both types of models have to be extended to capture the temporal constructs and other special features. Associated with a data model is a query language. The language should support the constructs needed to manipulate the multimedia database. For example, one may need to query to play frames 500 to 1000 of a video script. Query languages are discussed in another chapter. Languages such as SQL are being extended for MM-DBMSs.

Note that we have used a rather coarse-grained representation of multimedia databases with both object models and object-relational models. With the models we have used, we cannot represent the fact that Object A and Object B have to be played together. One needs to augment the models with rules to represent the constraints. We will discuss rule representation in hypersemantic data models in the next subsection.

In summary, several efforts are under way to develop appropriate data models for MM-DBMSs. Standards are also being developed. This is an area that has become mature within the last few years. For a discussion of object models, refer to Banerjee et al.;[14] for a discussion of relational models, refer to Codd.[28] Object-relational models are discussed in a special issue of *Communications of the ACM*.[1]

3.2.3 HYPERSEMANTIC DATA MODELS

Hypersemantic data models are essentially semantic data models with support for representing constraints and rules. They were studied by Trueblood and Potter[130] and have been extended for various secure applications by Binn.[18] One can consider both object and object-relational data models to be semantic models that capture the semantics of an application. It was noted in the previous subsection that with both object and object-relational data models, one cannot represent constraints in a natural manner. Although one could enforce them as part of methods, it is better to express them in a more natural way, such as with rules. Hypersemantic models enable the representation of constraints. Note that there is no standard hypersemantic data model. One can extend any type of semantic model such as semantic nets as well as objects with rules in a hypersemantic data model.

Figure 3.5 illustrates an example of using a hypersemantic data model to represent the objects of Figure 3.2. In addition to representing the objects, there are additional constraints such as (1) Object A has to be played before Object B, and (2) Object B has to be played before January 1, 2001. That is, temporal constraints are also enforced in this model.

3.3 INFORMATION MODELING

Information models are essentially data models combined with process models, as illustrated in Figure 3.6. While data models represent data, process models represent activities (see, for example, the discussion in Benyon[16]). Process models are essentially models that represent the business processes. Information models, therefore, represent the application's active as well as passive objects. Examples of information models are the object models for design and analysis. These include models such as the unified modeling language (UML) and its previous versions, including object modeling technique (OMT). These models can be found in Fowler et al.[44] and Rumbaugh et al.[105]

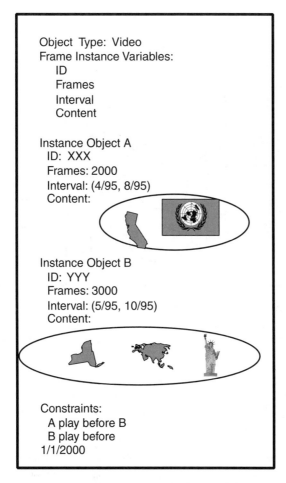

FIGURE 3.5 Data representation with hypersemantic data model.

FIGURE 3.6 Information models.

Thuraisingham[122] illustrates how multimedia applications may be modeled using OMT. OMT essentially has three models: an object model to represent the passive entities, a dynamic model to essentially represent the workflow, and a procedural model to represent the methods. Examples of multimedia applications include producing a video, preparing a multimedia presentation, purchasing a video on the Internet, and collaboratively analyzing video and audio data to make decisions. All of these applications have to access multimedia data.

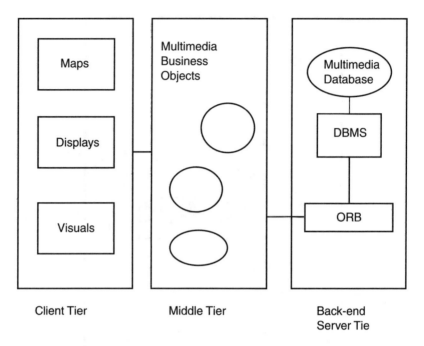

FIGURE 3.7 Three-tier representation.

Figure 3.7 illustrates the three-tier architecture for multimedia applications. The client tier is responsible for presenting the multimedia data. All of the workflow, such as the steps to produce a video, is represented by the business objects. The database stores the multimedia data. Note that this architecture was also discussed in Chapter 2.

Information models for multimedia applications are a rather new technology area. As we make more progress on multimedia database modeling as well as use multimedia data for applications such as e-commerce, we will see more and more information models for multimedia applications.

3.4 SUMMARY

This chapter discusses various aspects of data and information modeling for multimedia databases and applications. First, we provided an overview of data models. These include object-oriented data models, object-relational data models, and hyper-semantic data models. We then discussed information modeling.

As stressed in this chapter, a good data model is critical for multimedia applications. The data model should not only capture the complex data types but also the temporal relationships between the data objects. More details on data models can be found in Proceedings of the IEEE Multimedia Database Systems Workshop,[61-63] Nwosu et al.,[92] Prabhakaran,[102] Thuraisingham,[126] Vasirgiannis and Sellis,[133] and in a special issue of *IEEE Transactions on Knowledge and Data Engineering.*[129]

4 Metadata for Multimedia Databases

4.1 OVERVIEW

Metadata means many things to many people. Metadata started with a simple definition, i.e., data dictionary or schema. Initially, three-schema architecture was proposed for centralized databases. Then, for distributed databases, the schema architecture was extended to five layers. Additional considerations arose for heterogeneous and federated databases such as export schemas. Then came the Internet and the Web, and metadata included information about usage patterns, policies, and procedures as well as administration information.

For multimedia databases, metadata plays a key role. Metadata may include information about the multimedia databases, such as text, image, audio, and video data. This information could be of the form "Frames 2000 to 3000 contain information about the president's speech." In that case, the data is the president's speech. Metadata may also contain annotations for text, images, audio, and video. The four types of metadata for multimedia databases are illustrated in Figure 4.1.

This chapter explores the various types of metadata for multimedia databases. Metadata management aspects are also discussed. Specifically, Section 4.2 describes the various types of metadata, including metadata for text, images, video, and audio. Metadata management is the subject of Section 4.3. The chapter is summarized in Section 4.4.

4.2 TYPES OF METADATA

4.2.1 OVERVIEW

As illustrated in Figure 4.1, metadata may include text, image, audio, and video data. Metadata may also include combinations of data types such as audio and video or text and video. Metadata could also be characterized by whether it depends on the context or not. For example, in the case of text metadata, content-dependent metadata may include information that depends on the content, e.g., the keywords in the story are vehicle, employee, and license. Content-independent metadata may include information such as, "a book has chapters and paragraphs." Metadata may also be extracted from multimedia data and used to understand the multimedia data. This is illustrated in Figure 4.2.

This section describes the various types of metadata. Section 4.2.2 describes metadata for text. Metadata for image data is the subject of Section 4.2.3. Metadata for audio is the subject of Section 4.2.4. Section 4.2.5 describes metadata for audio

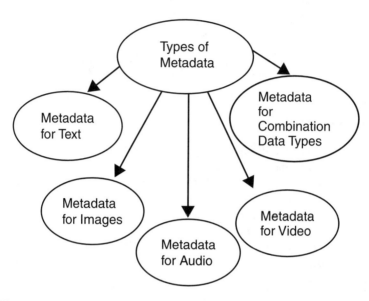

FIGURE 4.1 Types of metadata.

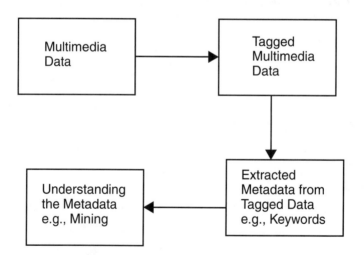

FIGURE 4.2 Extracting metadata from data.

data. Finally, Section 4.2.6 discusses metadata for combinations of data types and some other issues pertaining to metadata, such as ontologies, pedigree, and annotations. The use of XML (extensible markup language) for multimedia data will also be briefly addressed in this section. More details on XML and standards will be discussed in a later chapter (see also references 109, 140, and 141).

4.2.2 METADATA FOR TEXT

Metadata for text includes information about the text, which may be a book, newspaper, or essay. Content-independent metadata for text includes the type of text, the number of pages, the formatting of the text, and other information such as the number of chapters and paragraphs in each chapter. Content-dependent metadata may include information about the story and keywords.

One significant development for text data is the markup language called SGML (standard generalized markup language), which became a standard in the 1980s. With SGML, text data can be tagged so that metadata can be easily extracted. One can tag the places and people occurring in the text and also extract keywords from the tags. Developing appropriate tags is essential to this process. With the advent of the World Wide Web, SGML has evolved into XML.

Modeling metadata is also an important issue. One can place the tags in a relational database, or one can use an object model to represent the metadata. Special models such as those developed for information retrieval systems may be utilized for text data. One approach to understanding multimedia data is to extract and then the metadata mine. For example, metadata for text may be placed in relational databases, and then those relational databases can be mined. Mining text will be discussed in Part II of this book.

Text may also be annotated. While annotations are more useful for images and video, large volumes of text may be annotated so that only essential information is included in the annotations. Especially with Web-based systems, one can follow the links and get the annotations for the particular text being reviewed. Annotations can be compared to footnotes. Annotations can also be regarded as a type of metadata.

Figure 4.3 illustrates the types of metadata for text that have been discussed here. These include developing tags for metadata and subsequently extracting key words, developing models for the metadata, mining metadata for text, and developing annotations.

4.2.3 METADATA FOR IMAGES

Metadata for images may include text data describing the images, or the metadata may be stored in relational databases describing various properties of the images. For example, consider a picture illustrating an ocean with palm trees and beach homes. The text metadata for this picture may be, "the picture X describes an ocean with palm trees and beach houses." Another example is illustrated in Figure 4.4, where text metadata is used to describe a map of the United States.

As in the case of text, metadata for images may be described as annotations. For example, when one browses the Web, annotations about the images may be available when one clicks on the images. In the case of a large number of images, metadata may include partial images. That is, images may be used to describe the metadata. Tagging techniques have also been studied to tag images and subsequently extract metadata from the tags. The extracted metadata may be analyzed to understand the images. Image mining is an example of an analysis technique.

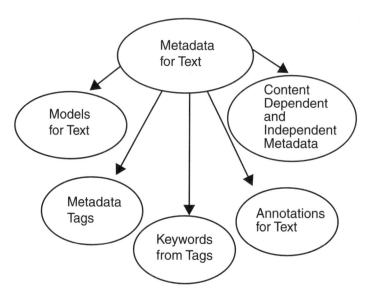

FIGURE 4.3 Metadata for text.

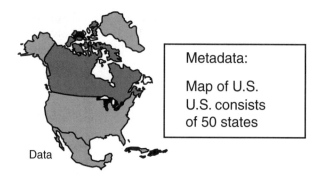

FIGURE 4.4 Image metadata.

Image metadata may also depend on the content or be content-independent. An example of content-dependent image metadata is, "the image is illustrating an ocean." Content-independent metadata may include information such as resolution and the number of pixels. Figure 4.5 illustrates the various types of image metadata.

4.2.4 METADATA FOR AUDIO DATA

Audio data is essentially speech data. For example, Figure 4.6 illustrates the audio frames containing the various speech topics by the president, specifically in this example, "Frames 1001 to 2000 contain speech on social security. Frames 2001 to 3000 contain speech on healthcare. Frames 3001 to 4000 contain speech on foreign policy." While the speech itself is audio data and is stored in audio databases, the text that describes this audio such as, "Frames 1001 to 2000 contain speech on social security" is the metadata.

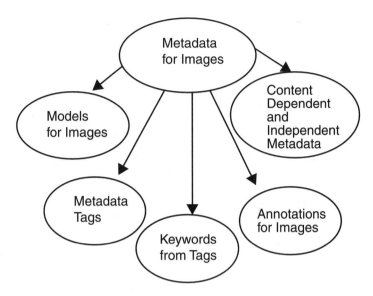

FIGURE 4.5 Types of image metadata.

Social Security Speech	Healthcare Speech	Foreign Policy Speech
Frames 1001-2000	Frames 2001-3000	Frames 3001-4000

FIGURE 4.6 Audio frames.

Audio metadata may not only be text but may also include audio clips. For example, one can use samples of speech from various people and then use voice recognition techniques to determine who is speaking. In that case, the samples of speech are the metadata. Parts of a speech can be used as the metadata.

As in the case of text and images, audio data may be tagged. Tagging audio data is a relatively new topic. One can also tag the text describing the audio and then extract and analyze metadata from the text. Figure 4.7 illustrates the various types of audio metadata.

4.2.5 METADATA FOR VIDEO DATA

Video data, like audio data, is continuous media. Note that text and images are still or noncontinuous data. In many ways, video and audio data are handled in a similar manner. For example, Figure 4.8 illustrates an example where "Frames 1001 to 2000 contain video clips about the president's visit to England. Frames 2001 to 3000

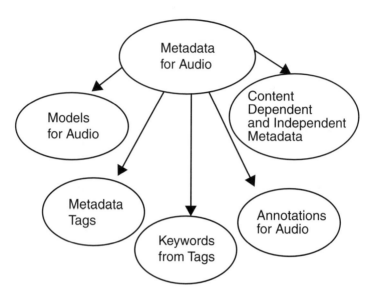

FIGURE 4.7 Types of audio metadata.

President's Visit to England	President's Visit to France	President's Visit to Japan
Frames 1001-2000	Frames 2001-3000	Frames 3001-4000

FIGURE 4.8 Video frames.

contain video clips about the president's visit to France. Frames 3001 to 4000 contain video clips about the president's visit to Japan." While the video clips (i.e., the data) are stored in the video database, information about the frames and what these frames contain is the metadata.

Video metadata may not just be text but can be images as well as video frames. For example, one can extract images from the video clips and store them as part of the metadata and later use this for video retrieval. An example would be, "find the portion of the video where A is shaking hands with B." The metadata may have images of A and B. Parts of the video can also be used as the metadata.

As in the case of text, images, and audio, video data may be tagged. Like audio, tagging video data is a relatively new topic. One may also tag the text describing the video and then extract and analyze metadata from the text. Figure 4.9 illustrates various types of video metadata.

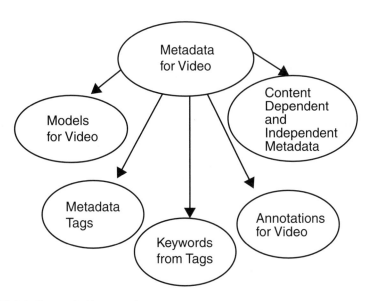

FIGURE 4.9 Types of video metadata.

4.2.6 OTHER ASPECTS

This section discusses five disparate aspects pertaining to metadata. One deals with metadata for combinations of data types, the second deals with ontologies, the third deals with annotations, the fourth deals with pedigree, and the fifth deals with XML for multimedia data.

First let us consider metadata for combinations of data types. Note that the previous subsection discussed metadata for individual data types. The challenge is handling the combinations of these individual data types. For example, in one case, the audio and video may be integrated, as shown in Figure 4.10, while in another case, the video and audio data may be in different databases. In the latter case, metadata may be used for the synchronized display of the two media, as illustrated in Figure 4.11. A schema architecture can also be developed to handle different media types, as illustrated in Figure 4.12.

Next, let us examine ontologies for multimedia data. Ontologies are essentially common representations for entities such as people, vehicles, tables, and stock quotes. Ontologies have been developed for various entities and are becoming essential for various applications. Ontologies are also developing for e-commerce applications. One of the areas being explored is the development of ontologies for multimedia data such as video objects, audio objects, text objects, and image objects. A hypothetical example is illustrated in Figure 4.13.

Our discussion of metadata for individual data types included annotations for metadata. For example, video and image data may be annotated with text. This text is part of the metadata. One challenge that arises is management of the annotations. That is, how are annotations extracted, queried, and updated, and how does one index the annotations and find the correlations between the annotations and the

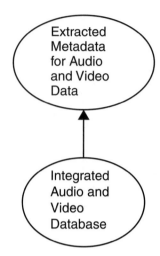

FIGURE 4.10 Metadata for an integrated database.

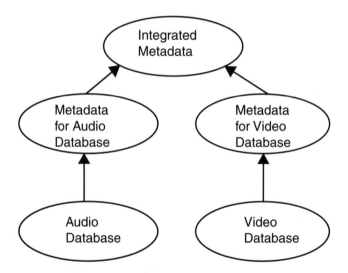

FIGURE 4.11 Metadata for a non-integrated database.

original data that is supposed to be annotated? Figure 4.14 illustrates the architecture for an annotation management system with functions for extraction, query, update, indexing, and even managing the metadata for the annotations.

Another closely related aspect for metadata is data quality and pedigree. How accurate is the data? What is the source of the data? This is important for multimedia data as well. For example, in the case of video data, who filmed the data? How reliable is the data? What about the resolution? What are the quality of service primitives for the display of the data? For example, under some conditions, it may be sufficient to have partial resolution, whereas in some other cases, one may need complete resolution. Data quality and pedigree issues are illustrated in Figure 4.15.

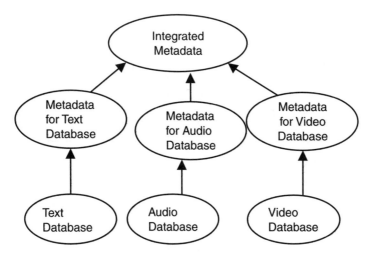

FIGURE 4.12 Schema architecture for multimedia databases.

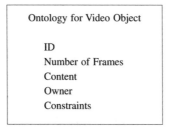

FIGURE 4.13 Example ontology for a video object.

In a later chapter, XML for multimedia data is discussed in more detail. XML is the extensible markup language for document exchange on the Web. With XML, one can develop a common representation for documents that can be understood by anyone. In its standard form, XML is primarily for text. XML is being extended to a variety of applications including multimedia and e-commerce. With multimedia XML and e-commerce XML, one can represent multimedia documents and documents for carryout e-commerce, respectively. Multimedia XML allows for the representation of not only text, but also video, audio, and image data. Figure 4.16 illustrates multimedia extensions for XML. More details are given by Decker et al.[34] and Hoschka.[56] Note that one type of extension to XML for multimedia data is called SMIL (synchronized multimedia integration language) and is discussed in Part III of this book (see also Decker et al.[34]).

4.3 METADATA MANAGEMENT

Section 4.2 described various types of metadata. This section describes the various issues concerning the management of metadata. We examine metadata management

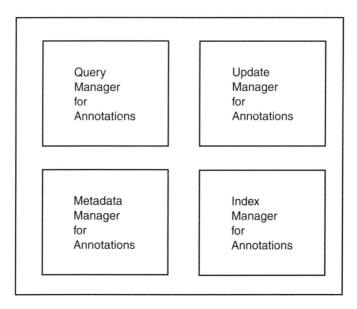

FIGURE 4.14 Data management system for managing annotations.

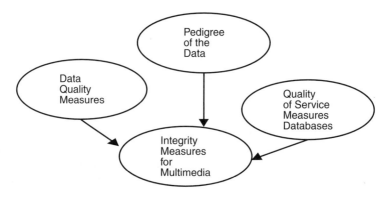

FIGURE 4.15 Multimedia data integrity.

for database systems, as discussed in Appendix B, and then discuss the enhancements needed for managing metadata for multimedia data.

Typically, the modules for managing metadata include extracting, querying, updating, indexing, and maintaining the security and integrity of the metadata. Metadata has an interface to the database system as well as a user interface. In fact, the database system may be a user of the metadata. Metadata management modules are illustrated in Figure 4.17.

For multimedia data, all of the functions illustrated in Figure 4.17 have to be carried out. In addition, there may be complex metadata, and therefore query, update, and indexing strategies may be more complex. For example, metadata and its annotations

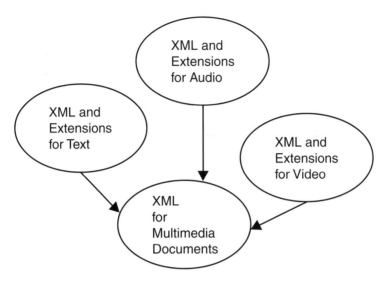

FIGURE 4.16 XML for multimedia data.

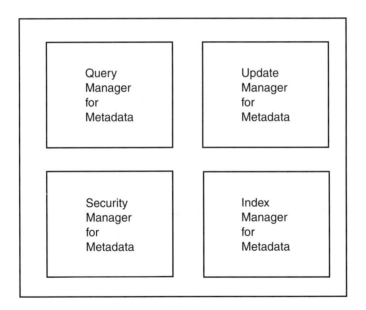

FIGURE 4.17 Metadata processor.

may need to be treated as transactions, as illustrated in Figure 4.18. What happens when metadata contains multimedia data? Metadata for video data may contain images. Another challenge is the management of the various annotations for multimedia data. Finally, multimedia data management modules may have to be integrated to manage distributed metadata, as shown in Figure 4.19.

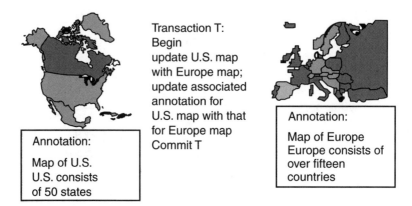

Transaction T:
Begin
update U.S. map
with Europe map;
update associated
annotation for
U.S. map with that
for Europe map
Commit T

Annotation:

Map of U.S.
U.S. consists
of 50 states

Annotation:

Map of Europe
Europe consists of
over fifteen
countries

FIGURE 4.18 Transactions and metadata.

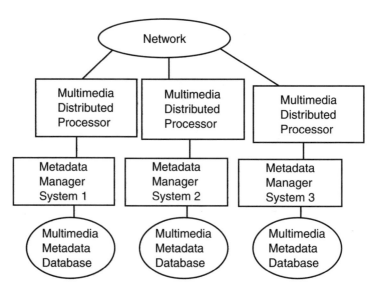

FIGURE 4.19 Distributed metadata.

4.4 SUMMARY

As mentioned earlier, metadata plays a key role in multimedia databases. Metadata
may include information about a multimedia database, including text, image, audio,
and video data. This information could be of the form, "Frames 2000 to 3000 contain
information about the president's speech." The data here is the president's speech.
Metadata may also contain annotations for text, images, audio, and video.

This chapter describes metadata for multimedia data. We started with a discussion of various types of metadata, in particular, metadata for text, images, audio,
and video. Both content-dependent as well as content-independent metadata were

discussed. We also discussed annotations for multimedia data. In addition, this chapter included a discussion of issues concerning extracting metadata from the data for text, video, audio, and images.

We next addressed various aspects of metadata. In particular, we discussed five disparate topics pertaining to metadata. One dealt with metadata for combinations of data types, the second with ontologies, the third with annotations, the fourth with pedigree, and the fifth with XML for multimedia data. As mentioned in this chapter, XML is an increasingly important development for specifying multimedia documents. Extensions to XML for multimedia data will be discussed in a later chapter. These extensions form the language SMIL.[56]

Finally, we discussed issues regarding metadata management, specifically, issues on querying the metadata, carrying out transactions on metadata, and distributing the metadata. Often, for multimedia databases, the metadata can be quite large. Therefore, efficient techniques are needed for managing the metadata.

Only recently has there been an explosion in research on metadata for multimedia database systems. Various conferences such as the IEEE Metadata Conference[84] has focused on metadata for multimedia data. As multimedia data management on the Web becomes common practice, there will be many more complexities to deal with involving the management of metadata for multimedia data.

5 Multimedia Query Processing

5.1 OVERVIEW

An MM-DBMS must support the basic DBMS functions. These include data representation, data manipulation (which includes query processing, editing, browsing, filtering, transaction management, and update processing), metadata management, storage management, and maintaining security and integrity. Chapter 3 discussed data representation and Chapter 4 discussed metadata management. This chapter discusses query processing, which includes a discussion of editing, browsing, filtering, transaction management, and update processing. Storage management will be the subject of Chapter 6. Security and integrity aspects will be addressed in a later chapter.

All of the database functions for an MM-DBMS are more complex because the data may either be structured or unstructured for multimedia databases. Furthermore, handling various data types such as audio and video is quite complex. In addition to these basic DBMS functions, an MM-DBMS must also support real-time processing to synchronize multimedia data types such as audio and video. Quality of service is an important aspect for an MM-DBMS. For example, in certain cases, high quality resolution for images may not always be necessary. Special user interfaces and multi-modal interfaces are also needed to support different media. We address some of these aspects in various parts of this book.

This section provides an overview of query processing for multimedia databases. Section 5.2 describes data manipulation, which includes query processing, editing, browsing, filtering, transaction management, and update processing. The main focus is on query processing aspects of data manipulation. Query language issues will be the subject of Section 5.3. The Chapter is summarized in Section 5.4.

5.2 DATA MANIPULATION FOR MULTIMEDIA DATABASES

5.2.1 DATA MANIPULATION FUNCTIONS

5.2.1.1 Overview

While query processing is the major focus of this chapter, data manipulation, which is much broader than query processing and includes editing as well as transaction management, is discussed here. Therefore, we first provide an overview of data manipulation and then discuss query processing.

As mentioned earlier, data manipulation involves various aspects. Support for querying, browsing, and filtering the data is essential for multimedia databases. In addition to querying the data, one may also want to edit the data. Query processing

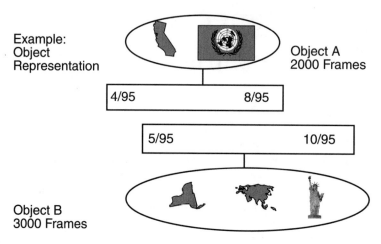

FIGURE 5.1 Data representation.

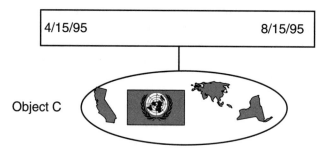

FIGURE 5.2 Example object editing.

is discussed in Section 5.2.2. This section provides an overview of some of the other aspects of data manipulation including a discussion of functions such as editing, browsing, filtering, transaction management, and update processing.

5.2.1.2 Object Editing

Consider editing multimedia objects. For example, two objects may be merged to form a third object. One can project an object to form a smaller object. As an example, objects may be merged based on time intervals, and an object may be projected based on time intervals. Objects may also be updated in whole or in part. Object editing, where the two objects illustrated in Figure 5.1 are merged over time intervals, is illustrated in Figure 5.2. Note that the objects in Figure 5.1 were also discussed in Chapter 3.

5.2.1.3 Browsing

Browsing multimedia data is essentially carried out by a hypermedia database management system. Hypermedia database systems were addressed in Chapter 2.

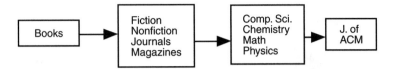

FIGURE 5.3 Browsing and linking.

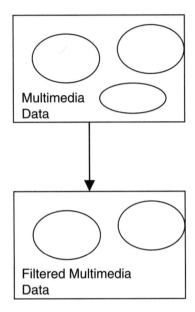

FIGURE 5.4 Filtering multimedia data.

We revisit them in Figure 5.3. The multimedia data is presented in terms of nodes and links. One traverses the links to reach the nodes and clicks on the links to get the relevant multimedia data.

5.2.1.4 Filtering

Filtering is the process of removing unnecessary material from data. This occurs quite often in video data where material inappropriate for children may be removed from a video clip. This means that the video clips have to be filtered and the filtered data displayed to the users. Figure 5.4 illustrates filtering.

5.2.1.5 Transaction Management

This function is also an aspect of data manipulation as it involves querying and updating databases. There has been some discussion as to whether transaction management is needed in MM-DBMSs.[3] We feel that this is important because, in many cases, annotations may be associated with multimedia objects. For example, if one updates an image, its annotation must also be updated. Therefore, the two

Transaction T:
Begin
update U.S. map
with Europe map;
update associated
annotation for
U.S. map with that
for Europe map
Commit T

Annotation:

Map of U.S.
U.S. consists
of 50 states

Annotation:

Map of Europe
Europe consists of
over fifteen
countries

FIGURE 5.5 Transaction processing.

operations have to be carried out as part of a transaction. Figure 5.5 illustrates an example of transaction management for an MM-DBMS.

Unlike query processing, transaction management in an MM-DBMS is still a new area. Associated with transaction management are concurrency control and recovery. The issue is, what are the transaction models? Are there special concurrency control and recovery mechanisms? Much research is needed in this area.

5.2.1.6 Update Processing

Update processing is usually considered part of query processing or transaction management. Update processing is essentially updating the multimedia data and is often a single user update. An example of a request is, "update the text paragraph C in document A to paragraph B." There have been many discussions on updating video and audio data. That is, how do you update part of a video or parts of an image? Is it possible to delete part of an image and replace it with some other image? We do not have satisfactory answers to update processing for multimedia databases. This is also related to the difficulties in transactions processing in multimedia databases. For an illustration of update processing, see Figure 5.6.

5.2.2 Query Processing

First, let us examine query processing aspects and then discuss the issues for multimedia databases. Query operation is usually the most commonly used function in a database system. It should be possible for users to query the database and obtain answers to their queries. There are several aspects to query processing. First, a good query language is needed. Languages such as SQL (structured query language) are popular for relational databases and are being extended for other types of databases. The second aspect is techniques for query processing. Numerous algorithms have been proposed for query processing in general and for the join operation in particular (see also Kim et al.[77]). Also, different strategies are possible to execute a particular query. The costs for the various strategies are computed, and the one with the least cost is usually selected for processing. This process is called query optimization.

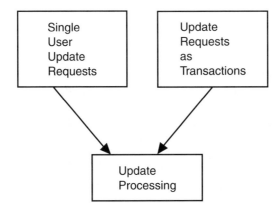

FIGURE 5.6 Example update processing.

Cost is generally determined by disk access. The goal is to minimize disk access in processing a query.

As stated earlier, users pose a query using a language. The constructs of the language have to be transformed into the constructs understood by the database system. This process is called query transformation. Query transformation is carried out in stages based on the various schemas. For example, a query based on external schema is transformed into a query on conceptual schema. This is then transformed into a query on physical schema. In general, rules used in the transformation process include the factoring of common subexpressions and pushing selections and projections down in the query tree as much as possible. If selections and projections are performed before the joins, then the cost of the joins can be considerably reduced.

Figure 5.7 illustrates the modules in query processing. The user interface manager accepts queries, parses the queries, and then gives them to the query transformer. The query transformer and query optimizer communicate with each other to produce

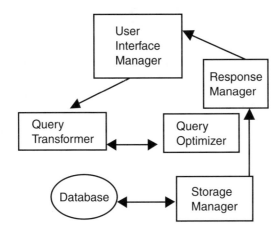

FIGURE 5.7 Query processor.

an execution strategy. The database is accessed through the storage manager. The response manager gives responses to the user.

Next, let us examine the impact of multimedia data on query processing. All of the modules for query processing are needed for multimedia databases. That is, the query transformer will take the query request and transform it into requests that can be understood by the system. The query optimizer will then optimize the query. Let us consider an example. Suppose a user requests to "play object A together with object B" and this query is in the form "play together objects A and B." This query will have to be parsed, and then an execution strategy has to be generated — in this case, to retrieve A and B and play them together. The system has to understand the meaning of playing two objects together. That is, special primitives for concurrent execution of the display of the two objects are needed. In addition, algorithms for synchronized display must also be implemented.

Other work in query processing for multimedia databases includes joins of multimedia data. Consider the object-relational data model discussed in Chapter 3 for multimedia databases. One can join the two tables on some common content. The query may be "find all the objects that have a common map of Europe in their content." Then, with the object relational representation, one could examine the contents of the objects and join those objects where their content has a map of Europe. Figure 5.8 illustrates such an example. Suppose we want to join the two objects illustrated in Figure 5.8a over the map of California. The object-relational representation of the objects is shown in Figure 5.8b. The result of the join on California is shown in Figure 5.8c. There is still lot of work to be done regarding merging, editing, displaying, and joining multimedia objects.

Note that while we have not shown the result in a single row, essentially what happens is that the object, frame, interval, and content for both objects will be displayed in one row for the resulting object.

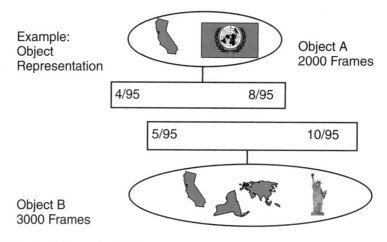

FIGURE 5.8a Objects to be joined.

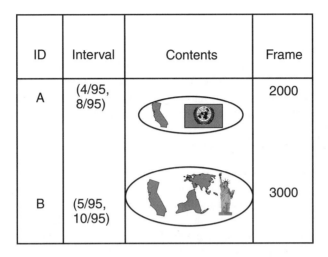

FIGURE 5.8b Object relation.

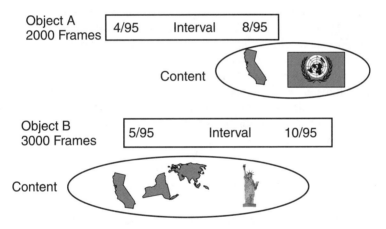

FIGURE 5.8c Result of the join.

5.3 QUERY LANGUAGE ISSUES

5.3.1 OVERVIEW

This section describes query language issues for multimedia database systems. The user poses the query in a language that is appropriate for an MM-DBMS. The user interface manager performs various functions and then gives the query to the query processor. The assembled responses are then presented to the user. Since multimedia data is complex, one needs interfaces for the different media types. For example, the interface for text may be quite different than the interface for video. Nevertheless, a common language to express queries is highly desirable.

Since SQL is a standard query language for databases, various extensions to SQL are being proposed for multimedia databases also. For example, in the case of multimedia databases, one may need to merge and edit the media objects as well as have constructs to play data together or play pieces of data before or after one another. A query language should support all these constructs.

Section 5.3.2 discusses SQL and the constructss needed for multimedia databases. User interface issues for query processing are the subject of Section 5.3.3.

5.3.2 SQL FOR MULTIMEDIA QUERIES

Various languages to manipulate relations have been proposed. Notable among these languages is the ANSI (American National Standards Institute) standard SQL. This language is used to access and manipulate data in relational databases.[110] SQL is widely accepted among database management system vendors and users. It supports schema definition, retrieval, data manipulation, schema manipulation, transaction management, integrity, and security. Other languages include the relational calculus first proposed in the Ingres project at University of California at Berkeley.[32]

SQL is used as both a data manipulation and data definition language. That is, SQL can be used to define the schemas and tables as well as manipulate the data for retrieval. The operations for data definition include create, delete, and insert. The operations for data manipulation include statements of the form, "select from table where some condition is satisfied." This select statement is used to retrieve data using just one relation or multiple relations.

With multimedia data, one may use SQL for selecting elements from an object-relation. For example, to retrieve the content portion of the multimedia object described in the previous section, an SQL statement would look as follows:

SELECT FROM TABLE

WHERE CONTENT = XXX

Here, table is the relation name, and content is the string XXX. However, with image and video data, we need to extend the language to support image and video content. That is, we need a way to specify the image, such as the U.S. map, or video, such as a particular film script. One possibility is to express the content with a string and then store the information that is the content in the metadata. That is, the string is the name of the frame, while the actual content may be the film. But there has to be a way to differentiate XXX and its content. Therefore, we need additional constructs, such as content in XXX. That is, the query may be of the following form:

SELECT FROM TABLE

WHERE CONTENT = IN XXX

We also need constructs to play-together, play-after and play-before. That is, to play video and audio data together, we may need a statement of the following form:

PLAY TOGETHER A AND B

WHERE A =

SELECT - - -

AND B =

SELECT - - -

That is, we need to play A and B together. To get A and B, we need select statements.

The above are just examples. A lot of work is going on to not only extend SQL but also to develop new multimedia query languages.[61,62,129] Note that SQL/MM contains the extensions that the standards groups have proposed for multimedia databases.

5.3.3 USER INTERFACE ISSUES

Note that a good user interface is needed to process the queries initially. These queries could be expressed via SQL, forms, pictures, and graphics. Therefore, the user interface should support multi-modal interfaces. Research on user interfaces and database management is proceeding almost independently. Only recently have visualization tools been integrated with database management systems. For multimedia database management, a variety of interfaces must be provided. These include interfaces for communicating with video, audio, and text databases. In addition, interfaces to support SQL extensions for multimedia data as well as ODMG standards are needed. Figure 5.9 illustrates multiple user interfaces for an MM-DBMS. The various components of a user interface manager are illustrated in Figure 5.10. This manager interfaces both to the user and the MM-DBMS. It parses the request so that the MM-DBMS can understand it. The parsed request is given to the query processor module discussed in an earlier section. The user interface manager is also

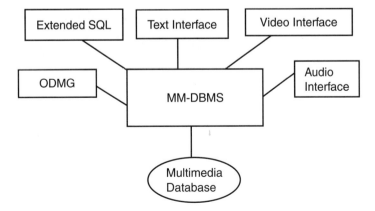

FIGURE 5.9 Multiple user interfaces.

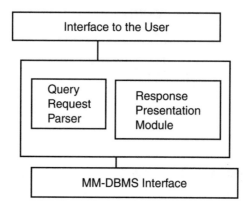

FIGURE 5.10 User interface manager.

responsible for presenting the response to the user. This is also illustrated in
Figure 5.10.

5.4 SUMMARY

This chapter has provided an overview of multimedia query processing. First, we
discussed data manipulation issues, in particular, browsing, editing, transaction processing, querying, and updating. We then discussed query processing at some length.
We gave examples of performing joins in multimedia databases, and we then discussed query language issues for multimedia database systems.

Much of the work on multimedia database systems has focused on query processing and data modeling. We need a good data model and query language as well
as strategies for query processing. This chapter has summarized the key issues. Next,
we need to determine techniques for efficient query processing. These techniques
are essentially good storage management techniques and are the subject of Chapter 6.

6 Multimedia Storage Management

6.1 OVERVIEW

While a user interacts with the multimedia system through queries, it is critical that the queries are processed efficiently. That is, efficient storage management techniques for retrieving data are needed so that users can get their responses in a timely manner.

This chapter deals with storage management for multimedia databases. With multimedia data, one may have to manage large quantities of data. For example, a video clip may consume gigabytes of data. Standard techniques for relational database management are not sufficient for multimedia databases. One needs special techniques for the efficient access of multimedia data.

Closely related to storage management are access methods and indexing. That is, the data may have to be indexed for efficient access. Furthermore, special methods are needed to access data. While access methods and indexing have been studied extensively, they are still not ready for application to multimedia databases.

This chapter describes access methods, indexing, storage management such as caching and synchronizing, and storage techniques. Section 6.2 discusses access methods and indexing as well as caching and synchronization. Various storage techniques are discussed in Section 6.3, and the chapter is summarized in Section 6.4.

6.2 ACCESS METHODS AND INDEXING

The storage manager is responsible for accessing the database. To improve the efficiency of query and update algorithms, appropriate access methods and index strategies have to be enforced. That is, in generating strategies for executing query and update requests, the access methods and index strategies that are used need to be taken into consideration. The access methods used to access the database depend on the indexing methods. Therefore, creating and maintaining appropriate index files is a major issue in database management systems. By using an appropriate indexing mechanism, the query processing algorithms may not have to search the entire database. Instead, the data to be retrieved could be accessed directly. Consequently, the retrieval algorithms are more efficient.

Extensive research has been performed to develop appropriate access methods and index strategies for relational database systems. Some examples of index strategies are B trees and hashing.[32] Current research is focusing on developing such mechanisms for object-oriented database systems with support for multimedia data.

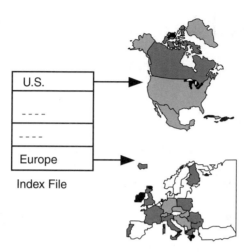

FIGURE 6.1 Indexing.

The major issues in storage management for multimedia databases include developing special index methods and access strategies for multimedia data types. Content-based data access is important for many multimedia applications. However, efficient techniques for content-based data access are still a challenge. Figure 6.1 illustrates an example for indexing. In that example, an index file is maintained on continents. There is a pointer to the corresponding map. It is assumed that many of the queries posed pertain to the continents of the world.

Next, we will discuss various aspects of indexing for different data types. First, let us consider text data. Some common indexing methods include keyword-based indexing. That is, one extracts keywords from the text and develops indexing for the keywords. That way, one can get to the part of the document that incorporates the keyword. For image data, indexing can be based on keywords such as the one illustrated in Figure 6.1, or one can develop an index containing images. For example, one may have a map of the U.S. as an index, and the corresponding link may point to all images that contain the U.S. within them. For video data, one could have text-based indexing or images for indexing. One can use images taken from the video and have the corresponding link point to all video frames with the image. Suppose we want to find all video clips that have "John" in them. We would form an index with John's image with the link pointing to all the clips that include John. For audio data, one could use text-based indexing or clips of audio for the index which then points to the audio frames of interest.

One could also develop indexes for the annotations. Since the annotations describe information about the multimedia data, if one can access the annotations, one can get the multimedia data. Various indexing techniques for multimedia data have been proposed.[61,62,92,126] Also, extensions to B trees and B + trees have been proposed for multimedia data. An example is the R tree discussed by Guttman.[51]

Other storage issues include caching data. How often should data be cached? Are there any special considerations for multimedia data? Are there special algorithms? Also, storage techniques for integrating different data types are needed. For

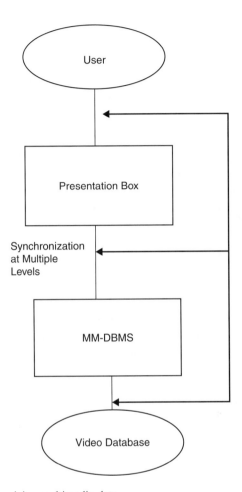

FIGURE 6.2 Synchronizing multimedia data.

example, a multimedia database may contain video, audio, and text databases instead of just one data type. The displays of these different data types have to be synchronized. Appropriate storage mechanisms are needed so that there is continuous display of the data. For example, consider the illustration in Figure 6.2. It is important that the display of multimedia data is synchronized with the retrieval of the data. This is especially true in the case of video on demand (VOD). Suppose we want to look at a film and we retrieve the video through the Internet or with the VOD boxes from our television sets. If the presentation is much quicker than the retrieval, there will be periods where we will have no display, which may not be acceptable to many viewers. In other words, we need continuous presentation of the film. In some cases, we can cache the film in order to get continuous display. It is impossible to cache all the films, and, therefore, we need efficient synchronization techniques. Typically, a user may specify the quality of service primitives, and the video should be presented according to the specifications. Video streaming has been a topic of much recent research. Furthermore, we now have special devices that consumers can purchase and attach to their television sets so that quality of service video on demand is possible.[102]

Storage management for multimedia databases is also an area that has received considerable attention. Several advances have been made during recent years.[82]

6.3 STORAGE METHODS

Various storage methods have been proposed in the literature. An excellent discussion of these methods is found in Prabhakaran.[102] In one method, called the single disk storage method, all of the multimedia data are stored on one disk, as illustrated in Figure 6.3. That is, the multimedia database has text, image, video, and audio data. The data may be stored contiguously to save space, or it may be scattered randomly. In the multiple disk storage method, objects are distributed across disks. Text data are stored on one disk, image data on a second disk, video data on a third disk, and audio data on a fourth disk. This is illustrated in Figure 6.4. In another type of

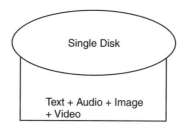

FIGURE 6.3 Single disk storage.

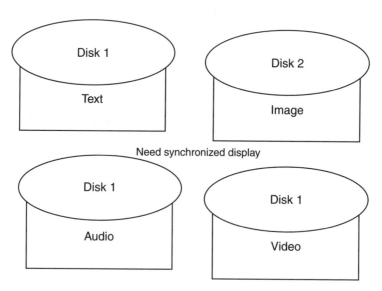

FIGURE 6.4 Multiple disk storage — approach 1.

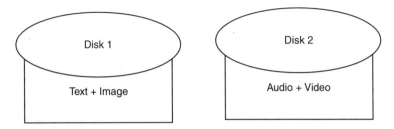

FIGURE 6.5 Multiple disk storage — approach 2.

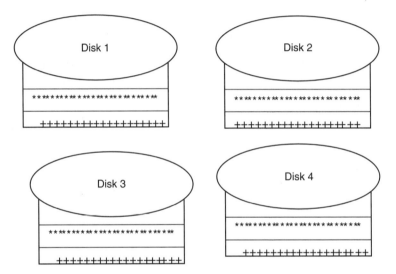

FIGURE 6.6 Disk striping.

multiple disk storage method, the multimedia data are distributed. For example, text and image data may be stored on one disk, while audio and video data may be stored on a second disk. This is illustrated in Figure 6.5. One may also use disk striping, where the multimedia objects are stored on more than one disk. That is, the objects are distributed across multiple disks, as shown in Figure 6.6.

For each of the techniques described here, efficient access methods and index strategies are necessary to efficiently access the data. For example, in the case of the single disk strategy, all of the objects are on one disk. This means that the system does not need to access multiple disks to answer a query. However, the system has to locate the object within the disk for display. If audio data is stored on one disk and video data on another, the scheduling techniques are needed for synchronized display of the data. In the case of disk striping, the system needs to scan multiple disks to compose the object.

There has also been extensive investigation of the use of RAID (redundant array of inexpensive disks) storage techniques as well as optical storage devices and jukeboxes for multimedia databases. In addition, one needs better techniques for handling very large databases since multimedia databases consume a lot of space. Some of the very large database issues are addressed by the Massive Digital Data Systems initiative.[82] There has also been an investigation of replication techniques to handle disk failures. Standards such as MPEG (moving picture expert group) and JPEG (joint photographic expert group) are being developed for compressing image, audio, and video data. Standards are discussed in Part III.

6.4 SUMMARY

This chapter briefly describes access methods, indexing, and storage strategies for multimedia databases. As mentioned in Chapter 5, a lot of work remains to be done for efficient query processing. Some key issues are addressed in this chapter; we began with a discussion of indexing for multimedia databases and gave some examples. We can index data with keywords, or we can use multimedia data such as images for indexing. Next, we discussed caching as well as synchronizing the display of multimedia data. Finally, we discussed storage strategies for multimedia data, i.e., single disk storage as well as multiple disk storage. The discussion was illustrated with examples.

Multimedia databases consume a lot of space. Therefore, access techniques developed for relational and even object database systems are not sufficient for multimedia database systems. Special techniques are needed not only for text, images, video, and audio data, but also for combinations of data types. Much work has been reported on storage and access methods for multimedia databases.[3,61-63,129]

7 Distributed Multimedia Database Systems

7.1 OVERVIEW

A distributed database system includes a distributed database management system (DDBMS), a distributed database, and a network for interconnection. The DDBMS manages the distributed database, which is data that is distributed across multiple databases. Distributed database system functions include distributed query management, distributed transaction processing, distributed metadata management, and enforcing security and integrity across multiple nodes. This chapter describes the impact of multimedia data on distributed database systems. That is, we discuss various aspects of distributed multimedia database systems (DMM-DBMS). For example, one node may have audio data and another node may have video data. In some cases, the multimedia database may also be distributed. We discuss concepts such as architecture and database design, and functions such as query processing, transaction management, metadata management, and security and integrity management for distributed multimedia database systems. We also address interoperability issues.

The organization of this chapter is as follows. Architectures for distributed multimedia database systems are the subject of Section 7.2. Distributed database design is discussed in Section 7.3. Query processing, transaction management, metadata management, and security and integrity issues are discussed in Sections 7.4, 7.5, 7.6, 7.7, and 7.8, respectively. Interoperability issues are the subject of Section 7.9. Warehousing multimedia databases is the Subject of 7.10. The role of multimedia networks is the subject of Section 7.11. Section 7.12 concludes the chapter. Numerous articles have been published on distributed database management systems. These include the work reported by Bell and Grimson,[15] Ceri and Pelugath,[20] and Oszu and Valdurez.[99] However, very few articles have appeared on distributed multimedia database systems. Some initial work was reported by Thuraisingham.[122] Further work was reported by Prabhakaran[102] as well as in some of the recent conference proceedings on multimedia database systems.[61-63]

Developing a DMM-DBMS requires integrating techniques from MM-DBMSs and DDBMSs. An overview of the concepts of DDBMS and MM-DBMS was provided by Thuraisingham.[126] A summary of DDBMSs is given in Appendix B of this book. Chapters 2 through 6 of this book addressed multimedia database system functions. This chapter identifies the issues involved in developing a DMM-DBMS. A lot of work needs to be done before we can provide solutions to developing a DMM-DBMS. In addition to the issues discussed here, there are several other areas that need attention, including multimedia database administration, fault tolerance, and performance analysis. A discussion of all of these topics is beyond the scope of this book.

7.2 ARCHITECTURE FOR DISTRIBUTED MULTIMEDIA DATABASE SYSTEMS

Our choice architecture for a distributed multimedia database system is essentially the architecture described in Appendix B for a distributed database. We have adapted this architecture for a multimedia database system. Figure 7.1 illustrates this architecture. We have chosen such an architecture because we can explain the concepts for both homogeneous and heterogeneous multimedia database systems based on this approach. In this architecture, the nodes are connected via a communication subsystem. Furthermore, as illustrated in Figure 7.1, each node has its own local MM-DBMS that is capable of handling the local applications. In addition, each node is also involved in at least one global application. That is, there is no centralized control in this architecture. As shown in Figure 7.1, the MM-DBMSs are connected through a component called the distributed multimedia processor (DMP).

The DMP is a critical component of the DMM-DBMS — it connects the different local MM-DBMSs. That is, each local MM-DBMS is augmented by a DMP. The modules of the DMP are illustrated in Figure 7.2. The different components are the distributed multimedia metadata manager (DMMM), the distributed query processor (DMQP), the distributed transaction manager (DMTM), the distributed security manager (DMSM), and the distributed integrity manager (DMIM). The DMMM manages the global multimedia metadata, which includes information on the schemas that describe the relations in the distributed database, the way the relations are fragmented, the locations of the fragments, and the constraints enforced. The DMQP

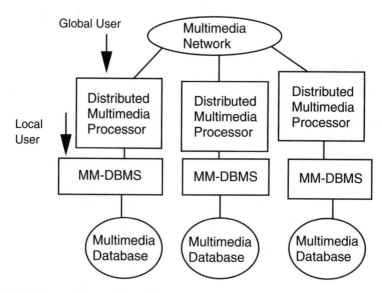

FIGURE 7.1 An architecture for a DMM-DBMS.

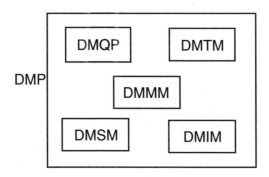

FIGURE 7.2 Modules of a DMP.

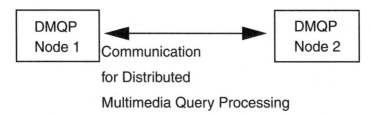

FIGURE 7.3 Peer-to-peer communication.

is responsible for distributed query processing; the DMTM is responsible for distributed transaction management; the DMSM is responsible for enforcing global security constraints; and the DMIM is responsible for maintaining integrity at the global level.

The DMQP, DMTM, DMSM, and DMIM all communicate with the DMMM for the metadata required to carry out their functions. The DMSM and DMIM also communicate with the DMQP and the DMTM because they process security and integrity constraints during query, update, and transaction execution. Our design of the DMP does not include a separate module for update processing. We assume that individual update requests are handled by the DMQP. Update requests specified as part of a transaction are handled by the DMTM. Since a transaction is a series of query and update requests, we assume that the DMQP is invoked by the DMTM in order to carry out the individual requests.

Note that the modules of the DMP communicate with their peers at the remote nodes. For example, the DMQP at node 1 communicates with the DMQP at node 2 for handling distributed queries. This is illustrated in Figure 7.3. As stated earlier, throughout this chapter we have assumed that the local MM-DBMSs are identical, i.e., the DMM-DBMS operates in a homogeneous environment.

7.3 DISTRIBITED MULTIMEDIA DATABASE DESIGN

Designing a distributed database includes several steps which are elaborated by Ceri and Pelagatti.[20] First of all, one needs to determine the schema at the global level.

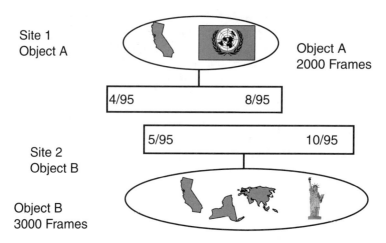

FIGURE 7.4 Distributed multimedia objects.

For example, the entire distributed database is viewed as a single entity, and the schemas are designed. In the case of a relational system, the schemas would describe the relations and attributes. The next step is to design the fragments. For example, in the case of a relational database, the fragments could partition a relation horizontally or vertically or both. The third step is to allocate the fragments to the various sites. It is at this stage that one determines whether the fragments are replicated or not. The last stage is to develop mappings between the global schema and the local schemas utilized by the databases. That is, if the global model is different than the local models, appropriate mappings must be developed. In a homogeneous environment, since we assume that all the local MM-DBMSs are identical, it is more meaningful to have global schemas utilizing the same representation model as the local systems. Finally, the physical schemas for the local MM-DBMSs are designed.

The issues for designing a distributed multimedia database are the same. A distributed multimedia database stored at two sites is illustrated in Figure 7.4, where the multimedia objects are stored at two sites. Figure 7.5 illustrates the object-relational representations for the distributed multimedia database. If a user queries to edit the objects to form a multimedia object over the interval 5/95 to 8/95, then the query processor will have to access both sites, retrieve the objects, and then merge the objects. The operations will become clear through the discussion of distributed query processing in the next section.

7.4 DISTRIBUTED MULTIMEDIA QUERY PROCESSING

There are two aspects to query processing in a DMM-DBMS: query transformation and query optimization. The query transformation process transforms a global query into equivalent fragment queries. This process is performed according to transformation rules and does not depend on the allocation of the fragments. Chapter 5 discussed some rules for query transformation in a centralized system. Additional rules for query transformation have been formulated for distributed database systems

Site 1

ID	Interval	Contents	Frames
A	(4/95, 8/95)	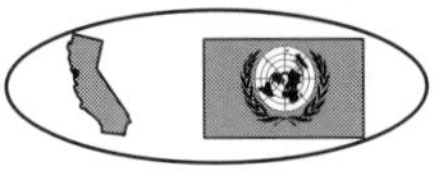	2000

Site 2

ID	Interval	Contents	Frames
B	(5/95, 10/95)	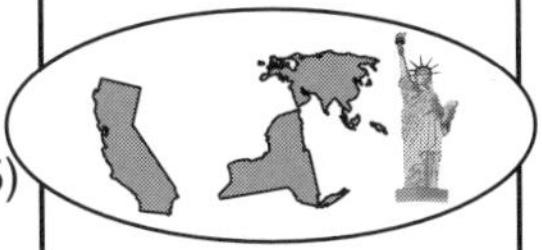	3000

FIGURE 7.5 Distributed multimedia object relations.

to take into account the data distribution. The current challenge is to determine the impact of multimedia data on query transformation. For example, if one database has audio data and another database has video data, then what are the challenges?

While query transformation focuses on modifying a query at the logical level, the query optimization process optimizes the query with respect to the cost of executing the query. That is, the alternate execution strategies produced by the query transformation process must be examined by the query optimizer in terms of the cost of executing them. The cost is usually determined by the amount of data transmitted. With multimedia data, one needs to determine a cost strategy for transmitting various data types such as text, audio, and video.

Figure 7.6 illustrates an architecture and strategy for distributed multimedia query processing. In that example, text, audio, and video data are distributed. The DMQP has to integrate all of the data before giving the data to the user.

7.5 DISTRIBUTED MULTIMEDIA TRANSACTION MANAGEMENT

Transaction management in a DDBMS involves the handling of distributed transactions. A distributed transaction refers to a transaction that executes at multiple sites. The portion of the transaction that executes at a particular site is a subtransaction associated with that site. A coordinator controls the execution of the subtransactions.

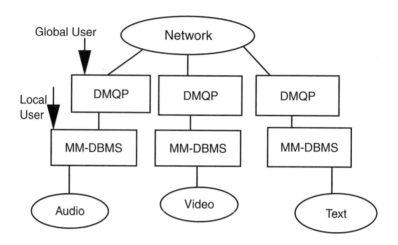

Query at site 1: Retrieve multimedia object with video, audio, and text data.
DMQP at site 1 will carry out the following:

Send Query 1 to MM-DBMS 1.
Send Query 2 to MM-DBMS 2 via DMQP at site.
Send Query 3 to MM-DBMS 3 via DMQP at site.
Send all results to DMQP at site 1.
Merge results at site 1.

Query 1: Retrieve audio from database at site 1.
Query 2: Retrieve video from database at site.
Query 3: Retrive text from database at site 3.

FIGURE 7.6 Distributed multimedia query processing.

Two types of log files are maintained at each node: one is the local log file which records the subtransaction execution, and the other is the global file which records the actions of the coordinator of all of the subtransactions. When a transaction commits, either all of the local subtransactions must commit or none should commit, in which case the transaction aborts. The technique used to ensure this requirement is the two-phase commit protocol.

If transactions are executed serially, there will be a performance bottleneck. Therefore, transactions usually execute concurrently. As a result, it must be ensured that the multiple transactions maintain the consistency of the distributed database. Concurrency control techniques ensure the consistency of the distributed database when transactions execute concurrently. There are additional problems in a DDBMS because of the replication of files. All of the copies of a file must be consistent. The techniques of locking, time-stamping, and validation have been proposed for concurrency control in a DDBMS. An additional problem of transaction management in a distributed environment is ensuring the consistency of the data in the presence of network partitions. Network partitions occur when certain nodes and/or links fail.

Let us now examine the impact of multimedia data on distributed transactions. Chapter 5 briefly addressed transaction management for multimedia database systems

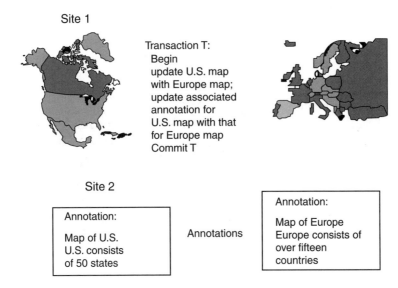

FIGURE 7.7 Distributed multimedia transaction processing.

in the discussion of data manipulation. One issue is whether the notion of a transaction makes sense. We illustrated by example how multimedia objects and their annotations have to be updated through a transaction. One can then apply this same reasoning for a distributed environment. That is, one can have the multimedia object at site 1 and the annotation at site 2, as illustrated in Figure 7.7. In that case, we need to update both the multimedia object as well as its annotations. Each update will be a subtransaction at the site. Both subtransactions will have to execute or the transaction will have to be aborted. Concurrency control and recovery for multimedia database systems is still a research issue. We need to develop solutions for a centralized environment before we can handle a distributed environment.

7.6 DISTRIBUTED MULTIMEDIA
METADATA MANAGEMENT

Metadata in a distributed environment includes information about the relations, fragmentation of the relations, allocation of the relations, and the local schema information. It is usually maintained at two levels. One is the global metadata and the other is the local metadata. Local metadata is the metadata associated with the local DBMS. Only global metadata is discussed here.

Global metadata includes information on the global schema, fragmentation schema, allocation schema, and all the mappings. These schemas are used for the various DDBMS functions such as processing and transaction management. Constraints such as global integrity and security constraints are also part of the metadata. In addition to the schema and constraint information, the metadatabase may also contain information on statistics of the database. The statistics include profiles of the database, access patterns, and history information.

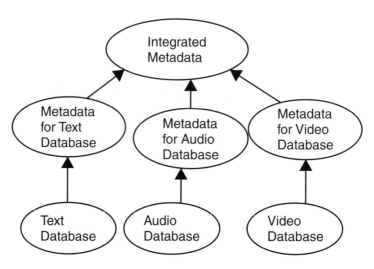

FIGURE 7.8 Distributed multimedia metadata processing.

Chapter 4 discussed metadata aspects for multimedia databases. As in the case of the centralized database system, efficient strategies for querying and updating the metadatabase are needed. In addition, in the case of a DDBMS, distributing the metadata is also an issue. For example, should the metadata be fully replicated, partially replicated, or not replicated at all?

For many applications, the distributed environment is becoming more and more complex. For example, for multimedia database management, metadata will include the annotations as well as information about the multimedia data such as which frames store what information. For hypermedia systems, metadata may also be used to guide the navigation/browsing process. That is, the metadata may be dynamic with information about the user's browsing patterns. Metadata may also include repository information needed to develop a global schema of the entire distributed system. That is, the various local schemas may have to be integrated to provide a global schema for users.

Figure 7.8 describes metadata architecture for distributed multimedia database systems. This can also be regarded as a schema architecture, described by Thuraisingham,[127] for distributed database systems. This architecture illustrates three databases that are distributed. The metadata has to be integrated to form the global metadata for the distributed multimedia databases. Note that this architecture was also described in Chapter 4.

7.7 DISTRIBUTED MULTIMEDIA DATABASE SECURITY

Security controls must ensure that users access only authorized information. These controls include discretionary security as well as multilevel security controls. Discretionary security mechanisms enforce rules that specify the types of access that users or groups of users have to the data. Multilevel security controls ensure that users cleared at different security levels access and share a distributed database in which the data are assigned different sensitivity levels without compromising security.

In a distributed environment, users need to be authenticated with respect to the local system as well as the global environment. One issue is whether the authentication mechanism should be centralized or distributed. If it is centralized, then the authenticator needs to have information about all of the users of the system. If the authenticator is distributed, then the various components of the authenticator need to communicate with each other to authenticate a user.

The access control rules enforced in a distributed environment may be distributed, centralized, or replicated. If the rules are centralized, then the central server needs to check all accesses to the database. If the rules are distributed, then appropriate rules need to be located and enforced for a particular access. Often, the rules associated with a particular database may also be stored at the same site. If the rules are replicated, each node can carry out the access control checks for the data it manages. Figures 7.9 and 7.10 illustrate architectures for secure distributed multimedia database management.

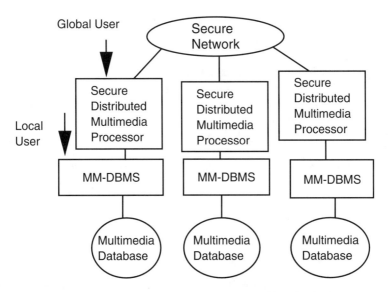

FIGURE 7.9 System architecture for a secure distributed multimedia database system.

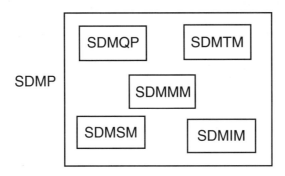

FIGURE 7.10 Modules of the secure distributed processor.

Figure 7.9 shows how an MM-DBMS is augmented by what we call an SDMP (secure distributed multimedia processor). The modules of an SDMP are illustrated in Figure 7.10. Several issues need to be investigated. First of all, an appropriate security policy has to be formulated. This policy will depend on the policies of the local MM-DBMS, the network, and the distributed processor. The algorithms for query, update, and transaction processing in a DDBMS have to be extended to handle security. Essentially, algorithms for MM-DBMSs have to be integrated with algorithms for DDBMSs to handle security policies. These algorithms are implemented by the SDMP.

7.8 DISTRIBUTED MULTIMEDIA DATABASE INTEGRITY

As stated by Thuraisingham,[126] there are various types of integrity enforcement mechanisms. These include traditional database integrity techniques such as concurrency control, recovery, and enforcing application-independent and application-specific integrity constraints. Integrity mechanisms also include techniques for determining the quality of the data.

Chapter 4 discussed some aspects of integrity. Those are re-enforced here. Data quality and pedigree are important for multimedia database management. For example, how accurate is the data? Where did the data come from? This is important for multimedia data as well. For example, in the case of video data, who filmed the data? How reliable is the data? What about the resolution? What are the quality of service primitives for the display of the data? For example, under some conditions, it may be sufficient to have partial resolution whereas for some other cases, one may need complete resolution. Data quality and pedigree issues are illustrated in Figure 7.11. A distributed system requires the development of techniques for integrity management in such an environment.

7.9 INTEROPERABILITY AND MIGRATION

Thuraisingham[126] discussed multimedia database systems in a section entitled Interoperation and Migration. Multimedia database systems are a special kind of database system that deals with heterogeneous data types. That is, as stated by Thuraisingham,[126] one aspect of heterogeneity focuses on handling different data types. For many new generation applications, there is a need to integrate different data types such as voice, text, video, and imagery. That is, a multimedia database system is needed to manage the different data types. This chapter describes some of the concepts of multimedia database systems.

As discussed by Thuraisingham,[126] there are other aspects of heterogeneity, such as handling different data models, query strategies, and security policies. These aspects also must be taken into consideration for multimedia database systems. Architecture for a heterogeneous multimedia database system is illustrated in Figure 7.12. As illustrated in that figure, the heterogeneous distributed processor (HDP) has to be adapted to process multimedia data. For example, the operations

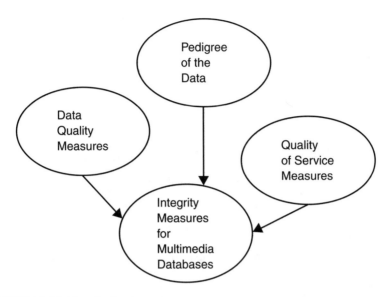

FIGURE 7.11 Multimedia data integrity.

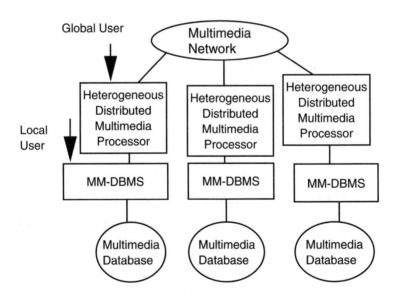

FIGURE 7.12 Heterogeneous multimedia database system.

of HDP, such as query processing, transactions management, metadata management, security management, and integrity management, have to be adapted to handle multimedia data. Essentially, we need to integrate techniques for distributed multimedia database management and heterogeneous database management. We call such an HDP a heterogeneous distributed multimedia processor (HDMP). Other aspects such as federated database management give rise to additional complexities.[106] A

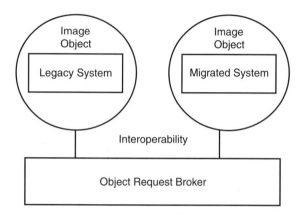

FIGURE 7.13 Migrating legacy multimedia databases.

discussion of federated multimedia database management is beyond the scope of this book.

Migrating legacy applications and databases also have an impact on multimedia database integration. Typically, a multimedia database environment may include legacy databases as well as some of the next generation databases. In many cases, an organization may want to migrate the legacy database system to an architecture like the client-server architecture and still want the migrated system to be part of the multimedia environment. This means that the functions of the multimedia database system may be impacted due to this migration process.

Two candidate approaches have been proposed for migrating legacy systems: one is to do all of the migration at once, and the other is incremental migration. That is, as the legacy system gets migrated, the new parts have to interoperate with the old parts. Various issues and challenges to migration are discussed by Thuraisingham.[126] Figure 7.13 illustrates an incremental approach to migrating legacy multimedia databases through the use of object request brokers.

7.10   MULTIMEDIA DATA WAREHOUSING

Data warehousing is one of the key data management technologies to support data mining. Several organizations are building their own warehouses. Commercial database system vendors are marketing warehousing products. In addition, some companies specialize only in developing data warehouses. What, then, is a data warehouse? It is often cumbersome to access data from heterogeneous databases. Several processing modules need to cooperate with each other to process a query in a heterogeneous environment. Therefore, a data warehouse will bring together the essential data from the heterogeneous databases. This way, the users only need to query the warehouse.

As stated by Inmon,[71] data warehouses are subject-oriented. Their design depends, to a great extent, on the application utilizing them. They integrate diverse and possibly heterogeneous data sources. They are persistent, that is, the warehouse is very much like a database. Data warehouses vary with time because as the data

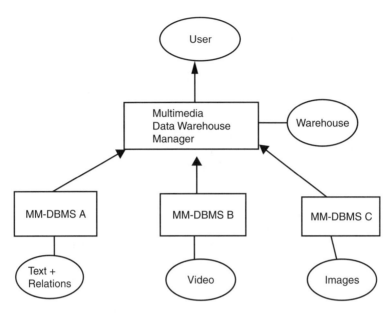

FIGURE 7.14 Multimedia data warehouse example.

sources from which the warehouses are built get updated, the changes are reflected in the warehouses. Essentially, data warehouses provide support to decision support functions of an enterprise or an organization. For example, while the data sources may have the raw data, the data warehouse may have correlated data, summary reports, and aggregate functions applied to the raw data.

Multimedia databases will also have to be integrated to develop a warehouse. For example, text and image data may be integrated with relational databases to form a warehouse. Figure 7.14 illustrates a data warehouse for multimedia data. The data sources are managed by database systems A, B, and C. The information in these databases is merged and put into a warehouse. There are various ways to merge the information. One is to simply replicate the databases, which does not have any advantage over accessing the heterogeneous databases. The second way is to replicate the information but remove any inconsistencies and redundancies. That has some advantages, because it is important to provide a consistent picture of the databases. The third approach is to select a subset of the information from the databases and place it in the warehouse. There are several issues involved. How are the subsets selected? Are they selected at random or is some method used to select the data? For example, one can take every other frame in a video clip and store those frames in the warehouse. The fourth approach, which is a slight variation of the third approach, is to determine the types of queries that users would pose, then analyze the data and store only the data that is required by the user. This is called on-line analytical processing (OLAP), as opposed to on-line transaction processing (OLTP).

With a data warehouse, data may often be viewed differently by different applications. That is, the data is multidimensional. For example, the payroll department may want data to be in a certain format, while the project department may want data to be in a different format. The warehouse must provide support for such multidimensional data.

In integrating the data sources to form the warehouse, one challenge is to analyze the application and select appropriate data to be placed in the warehouse. At times, some computations may have to be performed so that only summaries and averages are stored in the data warehouse. Note that the warehouse does not always have all the information for a query. In that case, the warehouse may have to get the data from the heterogeneous data sources to complete the execution of the query. Another issue is what happens to the warehouse when the individual databases are updated. How are the updates translated to the warehouse? How can security be maintained? These are some of the issues undergoing investigation. Furthermore, the impact of multimedia databases in data warehousing is an area that is largely unexplored.

We are often asked questions about the difference and/or relationship between warehousing and mining. Note that while data warehousing formats the data and organizes the data to support management functions, data mining attempts to extract useful information as well as predict trends from the data. Figure 7.15 illustrates the relationship between data warehousing and data mining. Note that having a warehouse is not necessary for mining, as data mining can also be applied to databases.

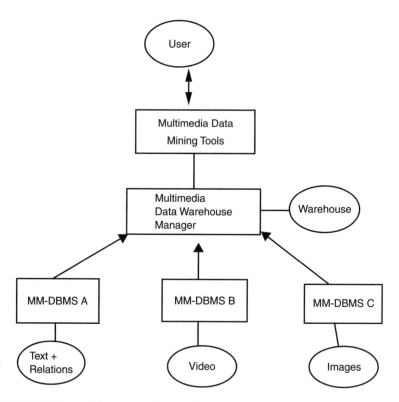

FIGURE 7.15 Data mining versus data warehousing.

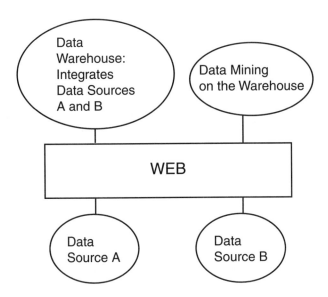

FIGURE 7.16 Data warehousing and mining on the Web.

However, a warehouse organizes the data in such a way as to facilitate mining; so, in many cases, it is highly desirable to have a data warehouse to perform mining. For many applications, building a good warehouse is one of the first steps toward multimedia data mining, which is discussed in Part II of this book.

Because we can expect large amounts of multimedia data to be managed on the Web, we need to build warehouses and mining tools to manage this data on the Web. Figure 7.16 illustrates warehousing and mining on the Web. We address multimedia on the Web in Part III of this book.

7.11 ROLE OF MULTIMEDIA NETWORKS FOR DISTRIBUTED MULTIMEDIA DATA MANAGEMENT

Up to this point, this chapter has focused entirely on data management issues for multimedia applications. While data is our main focus, it is also critical that appropriate networking support be provided for multimedia data management and data mining for an enterprise. Figure 7.17 illustrates the multimedia network that connects the multimedia server/DBMS to a client/application. We assume that the network provides all the necessary support. For completeness, we briefly review the requirements for multimedia networks. A more detailed discussion is given by Prabhakaran.[102] In addition, several networking aspects for multimedia applications and data management were addressed by the IEEE and ACM multimedia conferences.[6,61-63]

Let us examine some of the requirements. The network should be able to support the quality of service primitives. Typically, the user specifies what he wants. Those requirements have to be mapped down to the network. This would involve bandwidth as well as timing issues. That is, the network transmission algorithms should take into consideration the quality of service primitives that have been defined.

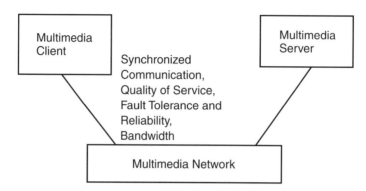

FIGURE 7.17 Multimedia networking for an enterprise.

The network should be reliable and fault tolerant. Since multimedia data is complex and large, constructing the multimedia objects in the presence of network and site failures may not be straightforward. The network must also ensure synchronized communication. For example, multimedia data cannot be displayed at one rate to the user and retrieved from the server at a much slower rate. If that was the case, the presentation would manifest as bursts of information. Therefore, the network must implement various scheduling algorithms. One can go down one level further and request the support that the operating system should provide. Essentially, we need some form of a real-time operating system for multimedia data, meaning that the network should also be able to handle timing constraints. This also imposes requirements on the middleware and the applications. We can learn a lot from real-time operating systems, networks, middleware, and databases for multimedia systems and applications.

Figure 7.18 illustrates three types of networks that are interconnected to form some sort of super network for a DMM-DBMS. Each circle is a node and can host a DBMS. These various DBMSs may be interconnected to form a DMM-DBMS. In network A, the nodes are connected through a single node. Network A can be regarded as a star network. In network B, all the nodes are connected to one another. The DDBMS based on our choice architecture is assumed to be hosted on such a network. In network C, some of the nodes are connected to one another. We also need to examine the use of such network configurations for multimedia applications and systems for an enterprise.

Note that we have provided a section on multimedia networks mainly for completion of our topic. Several comprehensive papers and books have been published on this topic that can provide the reader with for further information.[6,66]

7.12   SUMMARY

This chapter has described the impact of multimedia data on distributed database systems, that is, we discussed various aspects of distributed multimedia database systems.

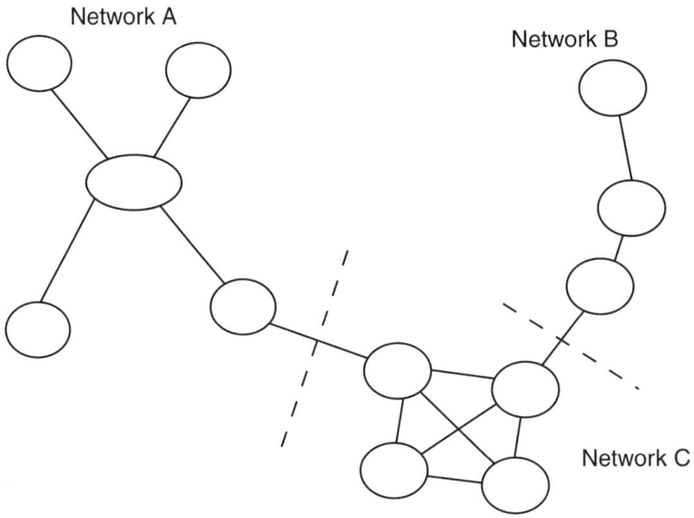

FIGURE 7.18 Networks for distributed databases.

We began this chapter with a discussion of distributed architectures and database design. We then focused on major functions such as distributed query processing, transaction management, metadata management, and security and integrity management. We then discussed interoperability, migration, and warehousing aspects for multimedia database management. As mentioned, much of the work on warehousing and mining has been carried out on relational databases. However, as we get more and more multimedia data on the Web, it is important to be able to warehouse the data and then apply the mining tools. Therefore, multimedia data warehouses will become popular in the future.

The field of distributed multimedia database management is still in its infancy. Only recently have commercial multimedia database systems become available in the marketplace. As we make progress in multimedia database management, data distribution, interoperability, and warehousing, we can expect to see more developments in this area. As mentioned earlier, some recent development on DMM-DBMS have been reported in Proceedings of the IEEE Multimedia Database Systems Workshop,[63] as well as by Prabhakaran[102] and Thuraisingham.[122]

Conclusion to Part I

Part I described various multimedia database management technologies. These include architectures, data modeling, query processing, metadata management, storage management, and data distribution. Essentially, we examined the functions of database systems and discussed the impact of managing multimedia data. Recent years have seen much progress in multimedia database systems management, and we now have commercial database system products.

Part II will show that we need good data to effectively mine data. Therefore, efficient multimedia database system management techniques are essential for effective data mining. That is, the discussions in Part I provide the background necessary to carry out data mining on multimedia data. With this background, we are now in the position to explore the various data mining technologies for multimedia databases in Part II.

Part II

Mining Multimedia Databases

INTRODUCTION

Part II describes multimedia data mining. Note that data mining is the process of posing queries to large databases and extracting information often previously unknown from the responses received using various techniques such as statistical reasoning and machine learning. While much of the work in data mining has focused on mining structured databases, it is also necessary to mine multimedia databases containing text, audio, images, and video. Multimedia data mining is a relatively new technology area.

Part II consists of Chapters 8 and 9. Chapter 8 describes multimedia data mining technologies and techniques. In particular, we examine the various technologies and techniques for data mining and discuss the impact of handling multimedia data. Chapter 9 focuses on mining data types such as text, images, audio, and video. We discuss the differences between text mining and text processing, image mining and image processing, video mining and video processing, and audio mining and audio processing. We also discuss various taxonomies for text, image, video, and audio mining, and we briefly address the issue of mining combinations of data types.

8 Technologies and Techniques for Multimedia Data Mining

8.1 OVERVIEW

Recently, there has been much interest in mining multimedia databases such as text, images, and video databases. As mentioned, many of the data mining tools work on relational databases. However, a considerable amount of data is now in multimedia format. There is a lot of text and image data on the Web, and news services provide a lot of video and audio data. This data has to be mined so that useful information can be extracted. One solution is to extract structured data from the multimedia databases and then mine the structured data using traditional data mining tools. Another solution is to develop mining tools to operate on the multimedia data directly. Technologies and techniques for multimedia data mining are the subject of this chapter. In particular, this chapter provides a general overview of multimedia data mining. Mining individual data types such as text, images, video, and audio is discussed in Chapter 9. Note that in order to mine multimedia data, we must mine combinations of two or more data types such as text and video, or text, video, and audio. However, this book mainly deals with one data type at a time because we first need techniques to mine the data belonging to the individual data types before mining multimedia data. In the future, we can expect tools for multimedia data mining to be developed.

The organization of this chapter is as follows. Section 8.2 provides some information on multimedia data mining technologies. Architectural support for multimedia data mining is discussed in Section 8.3. The process of multimedia data mining is the subject of Section 8.4. Section 8.5 discusses the outcomes of and techniques for multimedia data mining. Section 8.6 provides a summary of the chapter.

8.2 TECHNOLOGIES FOR MULTIMEDIA DATA MINING

8.2.1 OVERVIEW

This section provides a brief introduction to the technologies for multimedia data mining. One of the key technologies for data mining is multimedia database management and data warehousing. This is discussed at length in Part I of this book. However, there are several other technologies for multimedia data mining, Notable among these technologies are statistical methods and machine learning. Statistical methods have resulted in various statistical packages to compute sums, averages, and distributions. These packages are now being integrated with databases for mining. Machine learning is all about learning rules and patterns from data. One needs some amount of statistics to carry out machine learning. While statistical methods and machine learning are the two key components to data mining aside from data management, there are also some other technologies. These include visualization, parallel processing, and decision support. Visualization techniques help visualize the data so that data mining is facilitated. Parallel processing techniques help improve the performance of data mining. Decision support systems help prune the results and give only the essential results to help carry out management functions.

Sections 8.2.2 through 8.2.6 discuss the supporting technologies. We start with a discussion of multimedia data management and data warehousing. Statistical reasoning is addressed in Section 8.2.3. Machine learning is the subject of Section 8.2.4. Visualization issues are discussed in Section 8.2.5. Parallel processing is the subject of Section 8.2.6. Section 8.2.7 addresses decision support. Note that much of the information in this chapter is taken from one of our previous books.[127] However, this chapter focuses on the impact of managing multimedia data on data mining.

8.2.2 MULTIMEDIA DATABASE MANAGEMENT AND WAREHOUSING

As stressed repeatedly by Thuraisingham,[127] good data management is critical for mining. In the same way, good multimedia data management is critical for multimedia data mining. Therefore, we need multimedia database systems to effectively manage the data that can be mined. These systems could be data warehousing systems or the database systems discussed in Part I.

This chapter focuses on architecture, data modeling, database design, administration, and data warehousing issues for multimedia data mining. For example, should a data mining tool* be tightly integrated with a database system or should it be loosely integrated? What is the impact of data modeling on data mining? Can one design a database in such a way to facilitate mining? What is the impact on administration functions? Essentially, we discuss the impact of data mining on the various functions discussed in Part I. These include query processing, transaction management, storage management, metadata management, integrity and data quality, security, and fault tolerance.

Data mining techniques have been around for a while. That is, various statistical reasoning techniques, neural network-based techniques, and various other artificial intelligence techniques have been around for decades. So, why is data mining becoming so popular now? The main reason is that we now have the data to mine. Data is now being collected, organized, and structured, and database systems have played a major role in this. That is, with database systems, we can now represent, store, and retrieve the data and enforce features such as integrity and security.

Now that we have the data stored, how can we mine it? One approach is to augment a DBMS with a mining tool, as illustrated in Figure 8.1. One can buy a commercial MM-DBMS and a commercial mining tool that has interfaces built to

---

* We will also call such data mining tools data miners.

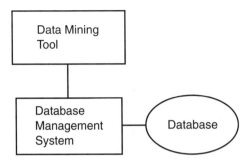

FIGURE 8.1 Loose integration between a DBMS and a data miner.

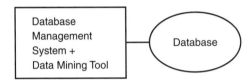

FIGURE 8.2 Tight integration between a DBMS and a data miner.

the DBMS and apply the tool to the data managed by the DBMS. That way, the mining tool does not have to be burdened with obtaining the data to be mined. While this approach has advantages and promotes open architectures, there are some drawbacks. There may be some performance problems when using a general purpose MM-DBMS for mining.

The other approach is a tight integration with mining tools, as shown in Figure 8.2. The database engine has mining tools incorporated within it. Such an MM-DBMS can be referred to as a mining MM-DBMS. In this approach, the various DBMS functions such as query processing and storage management are impacted by the mining techniques. For example, the optimization algorithms can be impacted by the mining techniques. There is a lot of research about the integration of mining into the MM-DBMS engine.[127] This research needs to be extended to mining multimedia data.

Mining a DBMS also would mean eliminating unnecessary functions of a DBMS and focusing on key features. Data mining is usually not conducted on transactional data but on decision support data. This data may not be updated often by transactions. Therefore, functions like transaction management can be removed from a mining DBMS, and one can focus instead on additional features such as providing data integrity and quality. Note that there are cases where transactional data has to be mined, for instance, mining credit card transactions.

In general, in the case of a mining MM-DBMS, every function may be impacted by mining. These functions include query processing, storage manager, transaction manager, metadata manager, and security and integrity manager. Therefore, we have added a data miner as part of a mining MM-DBMS, as illustrated in Figure 8.3.

Now let us focus on data modeling. The type of data model used may have some impact on mining. Chapter 3 discusses object models and object-relational data

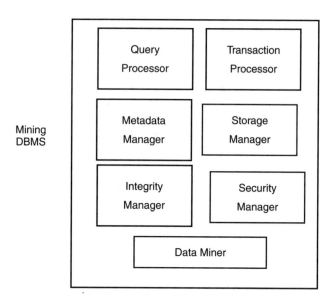

FIGURE 8.3 Functions of a mining DBMS.

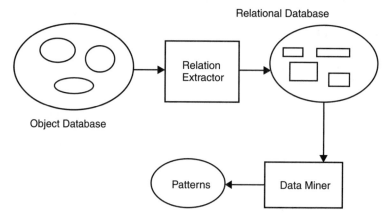

FIGURE 8.4 Mining object databases.

models for multimedia databases. If multimedia databases are to be mined, techniques for mining object and object-relational databases need to be developed.
Figure 8.4 illustrates an example of mining an object database where the relationships between the objects are first extracted and stored in a relational database, and
then the mining tools are applied to the relational database. This point will be
revisited again in the next chapter.

Database design plays a major role in mining. For example, in the case of data
warehousing, various approaches have been proposed to model and subsequently design
a warehouse. These include multidimensional data models and on-line analytical

processing models. Various schemas, such as the star schema, have been proposed for data warehousing. As mentioned, organizing data effectively is critical for mining. Therefore, such models and schemas are also important for mining.

Database administration is also impacted by mining. When integrating mining with an MM-DBMS, many questions arise, including how often should the data in the databases be mined? Can mining be used to analyze the audit data? If the data are updated frequently, how is mining impacted? We can expect answers to these questions as more information is obtained about integrating mining with the MM-DBMS functions.

Let us now examine the impact of multimedia data mining on the functions of an MM-DBMS, especially in the case of tight integration between the DBMS and the data miner. In that case, there is an impact on the various database systems functions. For example, consider query processing. There are efforts to examine query languages such as SQL and determine if extensions are needed to support mining.[5] If there are additional constructs and queries that are complex, the query optimizer has to be adapted to handle such cases. Closely related to query optimization are efficient storage structures, indexes, and access methods. Special mechanisms may be needed to support data mining in the query process.

In the case of transaction management, as mentioned earlier, mining may have little impact, since mining is usually performed on decision support data and not on transactional data. However, there are cases where transactional data are analyzed for anomalies such as credit card and telephone card anomalies. Many people have been notified by their credit card or telephone companies about abnormal patterns in usage. Those patterns are usually detected by analyzing the transactional data. Such data could also be mined.

One can mine metadata to extract useful information in cases where the data cannot be analyzed. This may be the situation for unstructured data whose metadata may be structured. On the other hand, metadata can be a very useful resource for a data miner; it can give additional information to help with the mining process.

Security, integrity, data quality, and fault tolerance are affected by data mining. In the case of security, mining could pose a major threat to security and privacy. On the other hand, mining can be used to detect intrusions as well as analyze audit data. In the case of auditing, the data to be mined is a large quantity of audit data. One may apply data mining tools to detect abnormal patterns. For example, suppose an employee makes an excessive number of trips to a particular country and this fact is discovered by posing some queries. The next query to pose is whether the employee has associations with specific people from that country. If the answer to that is positive, then the employee's behavior is flagged. The use of data mining for analyzing audit databases is illustrated in Figure 8.5.

Note that data mining has many applications in intrusion detection and analyzing threat databases. One can use data mining to detect patterns of intrusions and threats. This is an emerging area called information assurance. Not only is it important to have quality data, it is also important to recover from faults, malicious or otherwise, and protect data from threats or intrusions. While research in this area is just beginning, we can expect to see great progress.

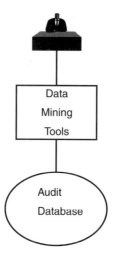

FIGURE 8.5 Mining an audit database.

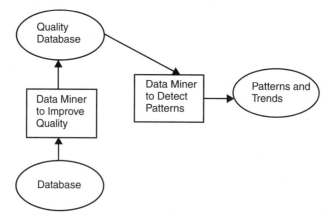

FIGURE 8.6 Data quality and data mining.

In the case of data quality and integrity, one could apply mining techniques to detect bad data and improve the quality of the data. Mining can also be used to analyze safety data for various systems such as air traffic control, nuclear, and weapons systems. This is illustrated in Figure 8.6.

Part I discussed multimedia data warehousing; we will revisit that topic here. A data warehouse assembles the data from heterogeneous databases so that users query only a single point. The responses that a user gets to a query depend on the contents of the data warehouse. The data warehouse, in general, does not attempt to extract information from the data in the warehouse. While data warehousing formats and organizes the data to support management functions, data mining attempts to extract useful information as well as predict trends from the data. Figure 8.7 illustrates the relationship between data warehousing and data mining. Note that having a warehouse

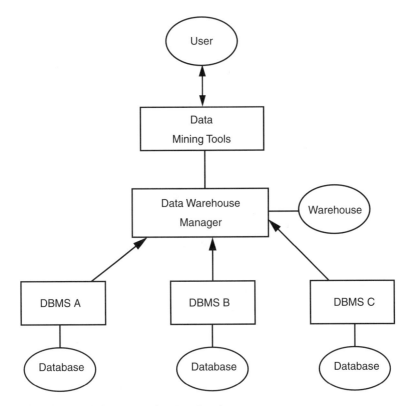

FIGURE 8.7 Data mining versus data warehousing.

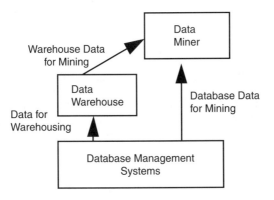

FIGURE 8.8 Database systems, data warehousing, and mining.

is not necessary for mining because data mining can also be applied to databases. However, a warehouse structures the data in such a way as to facilitate mining, so, in many cases, it is highly desirable to have a data warehouse to perform mining. The relationship between warehousing, mining, and database systems is illustrated in Figure 8.8.

Where does warehousing end and mining begin? For example, is there a clear difference between warehousing and mining? This answer is somewhat subjective. There are certain problems for which warehouses are the answer. Furthermore, warehouses have built in decision support capabilities. Some warehouses carry out predictions and trends. In those cases, warehouses perform some of the data mining functions. In general, we believe that in the case of a warehouse, answers to the query can be found in the database. The warehouse has to come up with query optimization and access techniques to get the answer. For example, consider questions like, "How many red cars did physicians buy in 1990 in New York?" The answer is in the database. However, for a question like, "How many red cars do you think physicians will buy in 2005 in New York?," the answer may not be in the database. Based on the buying patterns of physicians in New York and their salary projections, one could predict the answer to this question. In the case of video data, an example of a question would be, "Find the number of times the President of the United States appears with the Prime Minister of England."

Essentially, a warehouse organizes the data effectively so data can be mined. The question, then, is whether a warehouse is absolutely necessary to mine the data? The answer is that it is very good to have a warehouse, but having one is not necessary to mine data. A good MM-DBMS that manages a database effectively could also be used. Also, with a warehouse, one often does not have transactional data. Furthermore, the data may not be current, therefore, the results obtained from mining may not be current. If one needs up-to-date information, one can mine the database managed by a DBMS that also has transaction processing features. Mining data that keeps changing is a challenge. Typically, mining has been used for decision support data. Therefore, there are several issues that need further investigation before we can carry out what we call real-time data mining. For now, at least, we believe that having a good data warehouse is critical to perform good mining for decision support functions. Note that one can also have an integrated tool that carries out both data warehousing and data mining functions. Such a tool is called a data warehouse miner and is illustrated in Figure 8.9.

8.2.3 STATISTICAL REASONING

Statistical reasoning techniques and methods have been around for several decades. They were the sole means of analyzing data in the past. Numerous packages are now available to compute averages, sums, and various distributions for several applications. For example, the census bureau uses statistical analysis and methods to analyze the population in a country. More recently, statistical reasoning techniques are playing a major role in data mining. Some argue that the various statistical packages that have existed for quite a while are now being marketed as data mining products. This is not an issue for us because statistical reasoning plays just one role. Data mining requires the support of various other technologies including organizing and structuring the data. Many of the older statistical packages did not work with large relational databases. However, the packages that are being marketed today are integrated with various databases, as illustrated in Figure 8.10.

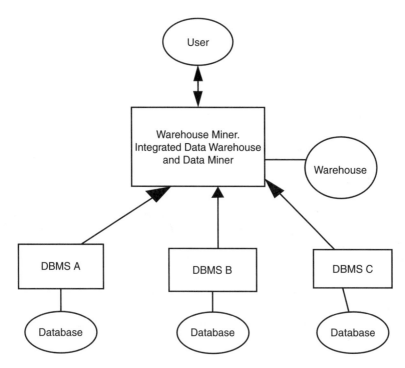

FIGURE 8.9 Integrated data warehousing and data mining.

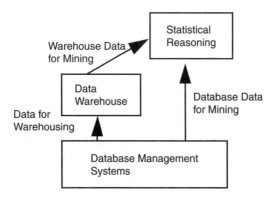

FIGURE 8.10 Statistical packages operating on databases.

As mentioned by Carbone,[19] the statistical techniques employed for data mining include those based on linear as well as nonlinear models. Linear regression techniques are employed for prediction, which is a data mining task that predicts variables from available data. For example, one could predict the salary of an employee in five years based on his current performance, his education, and market trends. Linear discriminate analysis techniques are used for classification. Classification is another data mining task, where an object is placed in a group based on some classifier. For example, John is placed in a class with others who earn more than $100,000. Nonlinear techniques are used to estimate values of new variables based on the data already available and characterized. Estimation is also a data mining task, where the various values are estimated based on available data. Sampling is a statistical technique used for data analysis. For example, in many cases, it will be impossible to analyze all of the data. Therefore, one draws samples, such as every nth row in a relation, and then forms and analyzes the sample. There has been some criticism of using samples for data mining.

In general, one cannot categorically state that linear models are used for classification while nonlinear models are used for estimation. There is no well defined theory in this area. The point is that statistics play a major role in data analysis. Even in machine learning, statistics play a key role. Unless one understands statistics and some operations research, it will be difficult to appreciate machine learning. Because of this, one cannot study data mining without a good knowledge of statistics. This section discusses statistical reasoning only very briefly. We refer the reader to the numerous texts on this subject for more in-depth information. One useful book is by DeGroot.[36] Mitchell's book on machine learning[85] contains a good introduction to the statistical terms and techniques needed for data mining. These include random variables, probability distribution, standard distribution, and variance. We now need to determine how statistical reasoning may be applied for multimedia databases.

8.2.4 MACHINE LEARNING

Machine learning involves learning rules from data. Essentially, machine learning techniques are those that are used for data mining. Therefore, the question is, while machine learning has been around for a while, what is new about its connection to data mining? Again, the answer is in the data. Only recently have the various machine learning techniques been applied to data in databases. These machine learning techniques are rapidly becoming data mining techniques.

Machine learning is all about making computers learn from experience. As Mitchell describes in his excellent text on machine learning,[85] machine learning involves learning from past experiences with respect to some performance measure. For example, in computer game applications, machine learning may mean learning to play a game of chess from past experiences, i.e., games that the machine plays against itself with respect to some performance measure such as winning a certain number of games.

Various techniques have been developed for machine learning. One such technique is concept learning, where one learns concepts from several training examples, neural networks, genetic algorithms, decision trees, and inductive logic programming. Each technique is essentially about learning from experience with respect to some performance measure. The various techniques are discussed in more detail later in this chapter. Several theoretical studies have also been conducted on machine learning. These studies attempt to determine the complexity of machine learning techniques.[85]

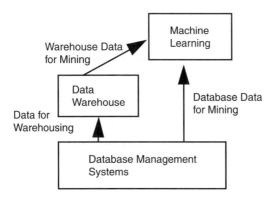

FIGURE 8.11 Machine learning and data mining.

Machine learning researchers have grouped some of the techniques into three categories. One is active learning, which deals with interaction and asking questions during learning, the second is learning from prior knowledge, and the third is learning incrementally. There is some overlap between the three methods. Various issues and challenges of machine learning and its relationships to data mining were addressed in a recent workshop on machine learning.[31] There is still a lot of research to be done in this area, especially on integrating machine learning with various data management techniques, as shown in Figure 8.11. Such research will significantly improve the whole area of data mining. Some interesting machine learning algorithms are given by Quinlan.[103]

8.2.5 VISUALIZATION

Visualization technologies graphically display data in databases. Much research has been conducted on visualization, and the field has advanced a great deal, especially with the advent of multimedia computing. For example, the data in a database could be rows and rows of numerical values. Visualization tools take the data and plot them in some form of a graph. The visualization models could be two-dimensional, three-dimensional, or more. Recently, several visualization tools have been developed to integrate with databases, and workshops have been devoted to this topic.[135] An example illustration of integration of a visualization package with a database system is shown in Figure 8.12.

More recently, there has been much discussion on using visualization for data mining. There has also been some discussion on using data mining to help the visualization process. However, when considering visualization as a supporting technology, it is the former approach that is getting considerable attention (see, for example, Grinstein and Thuraisingham[48]). As data mining techniques mature, it will become important to integrate them with visualization techniques. Figure 8.13 illustrates interactive data mining. In Figure 8.13, the database management system, visualization tool, and machine learning tool all interact with each other for data mining.

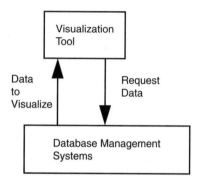

FIGURE 8.12 Database and visualization.

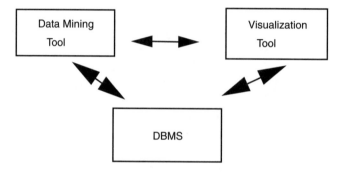

FIGURE 8.13 Interactive data mining.

Let us re-examine some of the issues of integrating data mining with visualization. There are four possible approaches. One is to use visualization techniques to present the results that are obtained from mining the data in the databases. These results may be in the form of clusters or they could specify correlations between the data in the databases. The second approach applies data mining techniques to visualization. The assumption there is that it is easier to apply data mining tools to data in a visual form. Therefore, rather than applying the data mining tools to large and complex databases, one captures some of the essential semantics visually and then applies the data mining tools. The third approach is to use visualization techniques to complement data mining techniques. For example, one may use data mining techniques to obtain correlations between data or detect patterns. However, visualization techniques may still be needed to obtain a better understanding of the data in the database. The fourth approach uses visualization techniques to steer the mining process.

In summary, visualization tools help interactive data mining, as illustrated in Figure 8.13. As shown in that figure, visualization tools can be used to visually display the responses from the database system directly so that the visual displays can be used by the data mining tool. On the other hand, the visualization tool can be used to visualize the results of the data mining tool directly. Little work has been done on integrating data mining and visualization tools. Some preliminary ideas were presented at the 1995 IEEE Databases and Visualization Workshop in Atlanta,

GA.[135] However, more progress was reported at the IEEE Databases and Visualization Workshop in Phoenix, AZ.[136] There is still much work to be done on this topic.

8.2.6 PARALLEL PROCESSING

Parallel processing is a subject that has been around for a while. The area has developed significantly from single processor to multiprocessor systems. Multiprocessor systems could be distributed systems or they could be centralized systems with shared-memory multiprocessors or shared-nothing multiprocessors. A lot of work has been done on using parallel architectures for database processing.[59] While considerable work has been carried out, these systems did not take off commercially until the development of data warehousing. Many data warehouses employ parallel processors to speed up query processing.

In a parallel database system, the various operations and functions are executed in parallel. While research on parallel database systems began in the 1970s, only recently are we seeing these systems used for commercial applications. This is partly due to the explosion of data warehousing and data mining technologies where performance of query algorithms is critical.

Let us consider a query operation which involves a JOIN operation between two object relations. If these object relations are to be sorted first on some attribute before the join, then the sorting can be done in parallel. We can take this a step further and execute a single join operation with multiple processors. Note that multiple tuples are involved in a join operation from both relations. Join operations between the tuples may be executed in parallel.

Many commercial database system vendors are now marketing parallel database management technology. This is an area we expect will grow significantly over the next decade. One of the major challenges therein is the scalability of various algorithms for functions such as data warehousing and data mining. We also need to determine how these algorithms may handle multimedia data.

Currently, parallel processing techniques are being examined for data mining. Many data mining techniques are computationally intensive. Appropriate hardware and software are needed to scale the data mining techniques. Database vendors are using parallel processing machines to carry out data mining. The data mining algorithms are put in parallel using various parallel processing techniques. This is illustrated in Figure 8.14.

Vendors of workstations are also interested in developing appropriate machines to facilitate data mining. This is an area of active research and development, and corporations such as Silicon Graphics and Thinking Machines have already developed products. We can expect to see a lot of progress in this area during the next few years.

8.2.7 DECISION SUPPORT

While data mining deals with discovering patterns from data, machine learning deals with learning from experiences in order to make predictions and perform analysis. Decision support systems are tools that managers use to make effective decisions. They are based on decision theory. One can consider data mining tools to be special

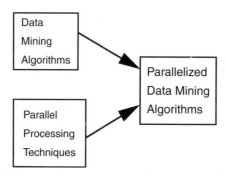

FIGURE 8.14 Parallel data mining.

kinds of decision support tools, as are tools based on machine learning, as well as tools for extracting data from data warehouses. Decision support tools belong to a broad category.[35]

In general, decision support tools could also be tools that remove unnecessary and irrelevant results obtained from data mining. They could also be tools such as spreadsheets, expert systems, hypertext systems, Web information management systems, and any other system that helps analysts and managers effectively manage large quantities of data and information. More recently, the field of knowledge management has emerged. Knowledge management deals with effectively managing an organization's data, information, and knowledge.[87,88] This includes storing the information, managing it, and developing tools to extract useful information. Some knowledge management tools also help in decision support.[132]

In summary, we believe that decision support is a technology that overlaps with data mining, data warehousing, knowledge management, machine learning, statistics, and other technologies that help manage an organization's knowledge and data. Figure 8.15 illustrates this concept. Figure 8.16 illustrates the relationship between data warehousing, database management, mining, and decision support.

8.3 ARCHITECTURAL SUPPORT FOR MULTIMEDIA DATA MINING

8.3.1 OVERVIEW

Like data management, data mining also needs architectural support. This section discusses various aspects of architectural support for data mining. Technology integration architecture is described first in Section 8.3.2. That section shows how various technologies have to be integrated for data mining. Functional architectural aspects are discussed in Section 8.3.3. Finally, system architectures including components are discussed in Section 8.3.4.

8.3.2 INTEGRATION WITH OTHER TECHNOLOGIES

One needs architectural support for integrating the technologies discussed in Section 8.2. Figure 8.17 shows a pyramid-like structure demonstrating how the various

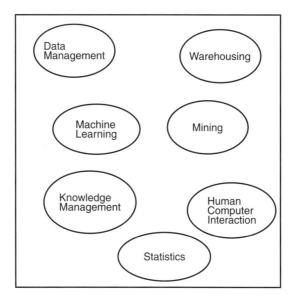

FIGURE 8.15 Decision support technologies.

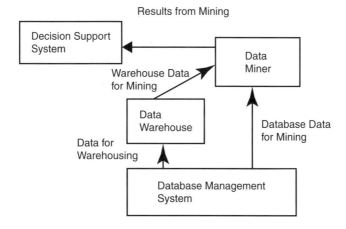

FIGURE 8.16 Decision support and multimedia data mining.

technologies fit with one another. As shown in Figure 8.17, communications and system level support are at the lowest level. Then comes middleware support, followed by database management and data warehousing. Above that are the various data mining technologies. Finally, the decision support systems that compile the results of data mining and help users effectively make decisions are at the top of the pyramid. Users could be managers, analysts, programmers, or any other user of information systems.

When one builds systems, the various technologies involved may not fit the pyramid identically, as we have shown. For example, one could skip the warehousing

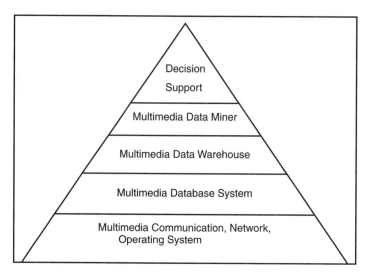

FIGURE 8.17 Pyramid for multimedia mining.

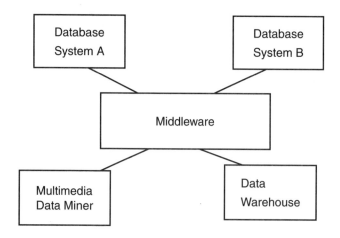

FIGURE 8.18 Revisiting the data mining architecture.

stage and go straight to mining. One of the key issues involved is the interfacing between the various systems. At present, there are no well defined standard interfaces except some of the standards and interface definition languages emerging from various groups such as the Object Management Group. However, as these technologies mature, one can expect standards to be developed for the interfaces.

The different multimedia data mining technologies have to work together. For example, one possibility is shown in Figure 8.18, where multiple databases are integrated through some middleware and subsequently form a data warehouse which is then mined. The data mining component is also integrated into this setting so that the databases are mined directly. Some of these issues will be discussed in the section on system architecture.

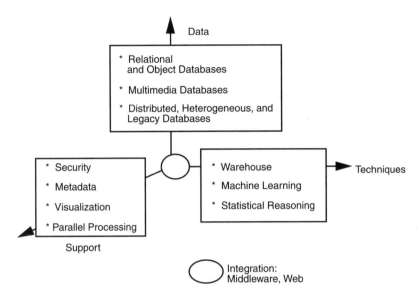

FIGURE 8.19 Three-dimensional view.

Figure 8.19 illustrates a three-dimensional view of data mining technologies. Central to this is the technology for integration. This is middleware technology such as distributed object management and Web technology for integration and access through the Web. On one plane are all the basic data technologies such as multimedia, relational, and object databases and distributed, heterogeneous, and legacy databases. On another plane are the data mining technologies. We have included warehousing and machine learning such as inductive logic programming and statistical reasoning here. The third plane consists of technologies such as parallel processing, visualization, metadata management, and secure access which are important for carrying out data mining.

8.3.3 FUNCTIONAL ARCHITECTURE

The steps for data mining will be elaborated in a later section. These steps also describe the functional components of data mining that are the subject of this section. Note that Part I discussed the functional components of a multimedia database management system. Now, a multimedia data miner could be part of an MM-DBMS; such a DBMS would be a mining MM-DBMS. This is illustrated in Figure 8.20. This approach considers data mining to be an extension to the query processor. That is, the query processor modules such as the query optimizer could be extended to handle data mining. Note that in this diagram, we have omitted the transaction manager, as data mining is used mostly for on-line analytical processing.

The question is, what are the components of the data miner? As illustrated in Figure 8.21, a data miner could have the following components: a learning from experience component that uses various training sets and learns various strategies, a data analyzer component which analyzes the data based on what it has learned, and a result-producing component that does classification, clustering, and other tasks

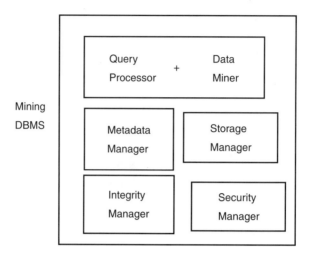

FIGURE 8.20 Data mining as part of query processing.

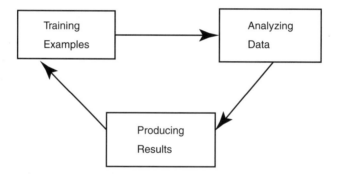

FIGURE 8.21 Data mining functions.

such as associations. There is interaction between all three components. For example, the component that produces results then feeds the results back to the training component to see if this component has to be adapted. The training component feeds information to the data analyzer. The data analyzer component feeds information to the result-producing component.

Note that we have not included components such as data preprocessor and result pruner into the data mining modules. These components are also necessary to complete the entire process. The data preprocessor formats the data; the data warehouse may also perform this function. The result pruner may extract only the useful information. This could also be carried out by a decision support system. All of these steps will be integrated into the data mining process discussed later.

8.3.4 SYSTEM ARCHITECTURE

Part I discussed various architectures for multimedia database management and interoperability. This section examines these architectures and discuss the impact of mining.

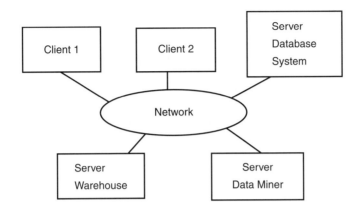

FIGURE 8.22 Client-server-based data mining.

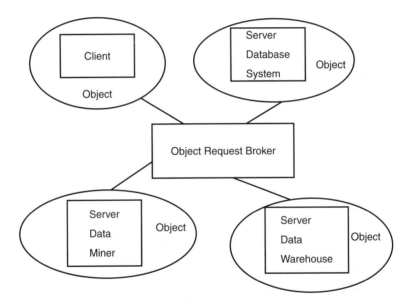

FIGURE 8.23 Data mining through an ORB.

Consider the architecture shown in Figure 8.22 In this example, the data miner could be used as a server, the database management system could be another server, and the data warehouse could be a third server. The client issues requests to the database system, warehouse, and miner, as illustrated in this figure.

One can also use an object request broker (ORB) for data mining. In that case, the data miner is encapsulated as an object. The database system and warehouse are also objects. This is illustrated in Figure 8.23. The challenge here is to define interface definition languages (IDLs) for the various objects.

Note that client-server technology does not develop algorithms for data management, warehousing, or mining. This means that algorithms are still needed for mining, warehousing, and database management. Client-server technology and, in

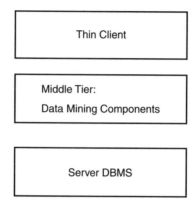

FIGURE 8.24 Three-tier architecture for data miner.

particular, distributed object management technology such as CORBA do facilitate interoperation between the different components. For example, the data miner, database system, and warehouse communicate with each other and with the clients through the ORB.

Note that three-tier architecture is becoming very popular (see the discussion in Thuraisingham[126]). In this architecture, the client is a thin client and does minimum processing, the server performs the database management functions, and the middle tier carries out various business processing functions. In the case of data mining, one could also utilize a three-tier architecture where the data miner is placed in the middle tier, as illustrated in Figure 8.24. The data miner can be developed as a collection of components that may be based on object technology. By developing data mining modules as a collection of components, one can develop generic tools and then customize them for specialized applications.

Another advantage of developing a data mining system as a collection of components is that one can purchase components from different vendors and assemble them together to form a system. Furthermore, components can be reused. Consider the following data mining modules: the data source integrator, the data miner, the results pruner, and the report generator. Each of these modules can be encapsulated as objects and ORBs can be used to integrate these different objects. As a result, a plug-and-play approach can be used to develop data mining tools. Figure 8.25 illustrates the encapsulation of the various data mining modules as objects. One can also decompose the data miner into multiple modules and encapsulate these modules as objects. For example, consider the modules of the data miner illustrated in Figure 8.3. These modules are part of the data miner module and can be encapsulated as objects and integrated through an ORB.

8.4 PROCESS OF MULTIMEDIA DATA MINING

8.4.1 OVERVIEW

Now that we have an understanding of what the multimedia data mining technologies are and how they contribute to data mining, let us discuss what data mining really

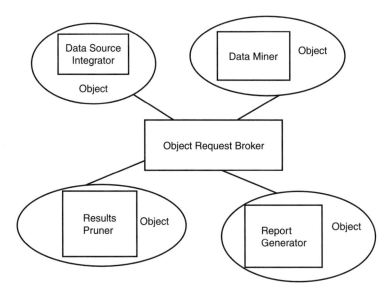

FIGURE 8.25 Encapsulating data mining modules as objects.

is and how mining is done. As mentioned in Appendix C, data mining is the process of posing queries and extracting previously unknown useful information, patterns, and trends from large quantities of data stored, possibly, in databases. That is, not only do we want to obtain patterns and trends, these patterns and trends must be useful, otherwise we can get irrelevant data that may turn out to be harmful or cause problems with the actions taken. For example, if an agency incorrectly finds that an employee has carried out fraudulent acts and the agency starts to investigate his behavior, and if this is known to the employee, then it could damage him. This is called a false positive. However, we also do not want results that are false negatives. That is, we do not want the data miner to return a result that the employee was well behaved when he is, in fact, a fraud. Therefore, data mining has serious implications, which is why it is critical that we have good data to mine and that we know the limitations of the data mining techniques.

We would like to stress to managers and project leaders not to rush into data mining. Data mining is not the answer to all problems, and it has sometimes been overemphasized. Carrying out the entire mining process is expensive and therefore has to be thought out clearly. In reality, mining is only a small step in the entire process. One needs to ask various questions such as, is there a need for mining? Do I have the right data in the right format? Do I have the right tools? More importantly, do I have the people to do the work? Do I have sufficient funds allocated to the project? All these questions have to be answered before embarking on a data mining project. Otherwise, the results may be extremely disappointing.

This chapter discusses data mining from start to finish without going into the technical details such as algorithms, approaches, and outcomes. Section 8.4.2 describes various example applications that might benefit from data mining. These are examples obtained from the various discussions, articles, and using common sense. Section 8.4.2 will give the reader a good sense of what data mining is all about. Section 8.4.3 discusses the reasons for performing data mining. For example, why is data mining such a buzzword now? What has changed about the world that makes data mining so useful now as opposed to twenty years ago when we already had computers? Section 8.4.4 discusses the steps involved in data mining. These steps include identifying and preparing the data, determining which data to mine, preparing the data to mine, carrying out the mining, pruning the results, taking actions, evaluating the actions, and determining when to mine next. Note that data mining is not a one time activity. An organization has to continually mine data because the data may have changed, the actions may not be beneficial, or the tools may have improved. Section 8.4.5 discusses some of the limitations and challenges of data mining, such as incorrect data, incomplete data, insufficient resources such as manpower, and inadequate tools. Some user interface aspects are discussed in Section 8.4.6.

8.4.2 SOME EXAMPLES

This section provides numerous examples that illustrate how data mining can be used. Some of these examples have been obtained from various papers and proceedings (see, for example, Grupe and Owrang[49]) and others from discussions. While much of the work in data mining supports marketing and sales, it is also useful in other areas. We keep multimedia data in mind when discussing these examples.

- A supermarket store analyzes the purchases made by various people and arranges the items on the shelves in such a way as to improve sales. Data here is usually in structured databases. However, there may be some text to mine.
- A credit bureau analyzes the credit history of various people and determines who is at risk and who is not. These histories may be in the form of text.
- An investigation agency analyzes the behavior patterns of people and determines who may be potential threats to protected information. The reports to be analyzed may be in the form of text.
- A pharmacy determines which physicians are likely to buy their products by analyzing the prescription patterns of physicians. While the data is usually in structured databases, it can also be in text databases.
- An insurance company determines potentially expensive patients by analyzing various patient records. These records may be in structured or text databases.
- An automobile sales company analyzes the buying patterns of people living in various locations and sends them brochures of cars that customers are likely to buy. Data may be in structured or text databases.
- An employment agency analyzes the various employment histories of employees and sends them information about potentially lucrative jobs. Data may be in text databases.
- An adversary uses data mining tools, gains access to unclassified databases, and somehow deduces information that is potentially classified. There may be a lot of text, images, video, and audio to mine.

- An educational institution analyzes student records to determine who is likely to attend their institution and sends those selected students promotional brochures. The records are likely to be in structured databases, although there may be text data.
- A nuclear weapons plant analyzes audit records of historical information and determines that there could be a potential nuclear disaster if certain precautions are not taken. There may be nuclear articles in text databases.
- A command and control agency analyzes the behavior patterns of adversaries and determines the weapons owned by those adversaries. Data may be in the form of text or even images.
- A marketing organization analyzes the buying patterns of various people and estimates the number of children they have as well as their income so that potentially useful marketing information can be obtained. While data is usually in structured databases, it is likely that there may be some text to mine.
- By analyzing the travel patterns of various groups of people, an investigation agency determines the associations between the various groups. Data could be in text or even image databases.
- By analyzing patient history and current medical conditions, physicians not only diagnose medical conditions but also predict potential problems that may occur. Data may be in image (like x-rays) and text databases.
- The income tax revenue office examines the tax returns of various groups of people and finds abnormal patterns and trends. Data may be in text form.
- An investigation agency analyzes the records of criminals and determines who is likely to commit terrorism and mass murders. Data may be in the form of text such as newspaper articles.

Note that we have selected various examples from all types of applications, including financial, intelligence, and medical, that carry out various activities such as marketing, diagnosis, correlations, and fault detection. These application areas are illustrated in Figure 8.26. All of these examples show that a great amount of data analysis is needed to obtain results and conclusions. This type of data analysis is usually referred to as mining. Note that mining is not always used in a positive manner to better human society. In many cases, it can be quite dangerous such as by compromising the privacy of individuals. This book provides information not only on the positive aspects but also on the negative aspects. In addition, we also discuss the difficulties and challenges involved in mining.

8.4.3 WHY DATA MINING?

Now that we have seen various applications that may use data mining, let us discuss why data mining has become so important now? We know that many of these problems have existed for several years. Furthermore, data has been around for centuries. The answer is that we are using new tools and techniques to solve problems in a new way. These problems have existed and people have worked on handling

FIGURE 8.26 Some data mining application areas.

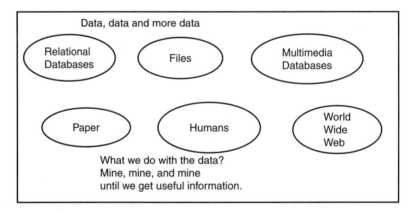

FIGURE 8.27 Why data mining?

them for years, for example, data analysis has been carried out for years in different ways, but only now do we call it data mining because of its improved methods and techniques (see Figure 8.27).

Although data has been around for a long time, it has primarily existed on paper and, in many cases, in the minds of people. Typically, clerks spent years recording data, and human analysts went through that data to detect various patterns. Then the whole field of statistics arose and provided a new way to analyze data. However, organizing data was still a big problem. Then, with computers and especially databases, we started storing data in computerized files and databases. That was the first big step toward data mining. Then came the area of artificial intelligence with new and improved searching and learning techniques. What has contributed most to data mining are the improved methods of storing and retrieving data, i.e., database management systems technology.

More recently, techniques and tools have been developed to focus on improving methods to capture the data and knowledge in organizations. This will be even better for data mining. There is still so much data out there that has not been captured. Furthermore, even if it could be captured, the existence of much of the data is not even known. So, knowledge management techniques will hopefully improve these deficiencies.[86]

Now we have large quantities of data computerized. The data could be in files, relational databases, multimedia databases, and even on the World Wide Web. Very sophisticated statistical analysis packages exist. Tools have been developed for machine learning. Parallel computing technology for improving performance is getting more mature. Visualization techniques improve our understanding of the data. Decision support tools are also maturing. So what better way to provide improved capabilities for analyzing data and predicting trends than integrating these various developments? Data mining has become a reality now, and, therefore, we are beginning to prepare for data mining technology. With respect to multimedia data mining, only recently are we getting a good handle on managing multimedia databases. Therefore, multimedia data mining technologies are still a few years into the future.

8.4.4 STEPS TO DATA MINING

We have given various examples of data mining and established a need for data mining. So, what are the steps to mining? Where do we start and where do we end? Various texts have discussed the steps of data mining (see, for example, Berry and Linoff[17]). Based on what we have read and from our experiences, the data mining steps, some of which are illustrated in Figure 8.28, are the following:

- Identifying the data
- Preparing the data
- Mining the data
- Getting useful results
- Identifying actions
- Implementing the actions
- Evaluating the benefits
- Determining what to do next
- Carrying out the next cycle

First, the data must be identified. As mentioned in the previous section, data can be all over the world, not just in one enterprise. Data can be distributed. It can be on paper and even in people's heads. Data can be in the form of text, images, video, and audio. We need to figure out what data we need and where we can find it, and then we can retrieve the data.

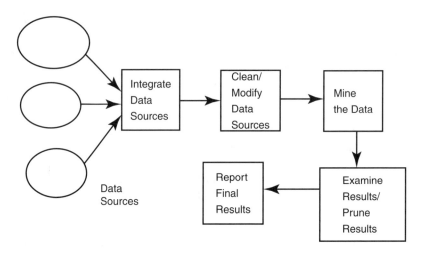

FIGURE 8.28 Steps of mining data.

Once we have the data, we need to prepare it. A lot of effort goes into this. Data has to be put in databases in the right format. We may need a data warehouse or database management system. This is by no means trivial, and this step is often underestimated. Preparing the data is one of the most difficult tasks in mining.

Once we have the data in the right format, we still need to clean the data, scrub unnecessary items, and keep only the data essential for mining. This is also not trivial. We now have the data we want to mine and we have seen how to go about mining. Next, what outcomes do we want? It is important to have some idea of the desired outcomes. Do we want the mining tools to find interesting patterns without letting the data miners know something about what we want? Then how do we go about getting what we want? Are there tools available? Must we build the appropriate tools? This step is very time-consuming.

Once we have determined the data and the tools to use, we let the tool begin to operate on the data. This is the easy part. The tool may produce a lot of data which may seem like a foreign language to many. What do we do with the patterns produced? Do we have an application specialist analyze the patterns? Are there analysts who can figure out what the data is all about? Do we have tools to analyze the results and determine useful patterns? That is, we need to figure out how to effectively prune the results and keep only the useful results. The pruning process is illustrated in Figure 8.29.

At this point, we think we have useful results. We now need to examine the results and identify actions that can be taken. For example, in a supermarket by analyzing the various purchases made, we decide that we need to put milk and moisturizing cream together. In other words, we identify the various actions we think will be beneficial and then discuss procedures to implement the actions.

Once the actions are implemented, we must wait to see the results, which could be immediate or take a very long time. Once we are in a position to determine the benefits and costs of our actions, we then re-evaluate the whole procedure. By then, the data may have changed. New tools may be available. We may have to do things

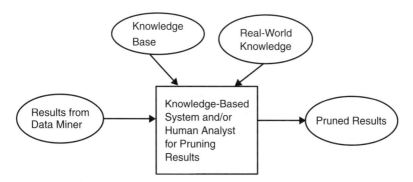

FIGURE 8.29 Pruning the results.

differently. So, we plan for the next mining cycle and determine how to go about mining the data.

Note that the above discussion does not bring the human element into the process. Humans play a major part. First of all, we need management buy-in. Therefore, we must be very careful not to oversell the project. We need to be realistic as to what mining can and cannot do for us. Once management is convinced, we need to determine whether we have the manpower. Do we train people or contract the work out? Also, we need to discuss these issues with our customers. Another issue is, what do we do about the tool developers? Do we use developers from inside or outside our corporation? The customer, contractor, and tool developer have to work very closely together to make mining a success.

If the project has failed, we cannot point a finger at one person or a group of people. This is still a new technology, so the likelihood of success is not high. We must learn from our experiences and talk to people who have had similar experiences. We must evaluate what can be done differently. In many cases, it might be good to create a small pilot project or prototype before performing full-scale mining.

8.4.5 CHALLENGES

Now we know the applications that need mining, why we perform mining, and the steps involved in mining. Next, we must examine the difficulties, both technical and otherwise. The nontechnical difficulties include a lack of management support and resources, such as trained individuals and low budgets. The technical challenges are many. Some of them are discussed below.

We have stressed that getting the right data in the right format is critical. But a lot of problems are involved. The data may not be accurate. What do we do then? How do we track down the source, as data may have passed many levels? This is one of the major challenges. Also, the data may be incomplete. There may be many missing values. How do we fill in the blanks? Additionally, the accuracy of the data may be uncertain. Do we track down the source? Therefore, missing, inaccurate, and uncertain data are major challenges.

Next, do we have the right tools? If not, what do we do? Do we get tools and adapt them or develop them from scratch? This is a big problem since the existing

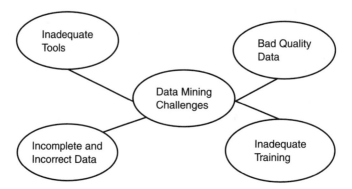

FIGURE 8.30 Some data mining challenges.

tools are still not mature. Another challenge is developing adaptive techniques. That is, many of the tools only do one type of data mining like classification or clustering. Can we develop tools that can adapt to a situation and carry out a particular type of mining? Can a tool use multiple mining techniques and handle different outcomes? There is research in this area, but we are a long way from robust commercial tools.

We have named a few of the challenges, and some of them are illustrated in Figure 8.30. The good news is that now there is a lot of research about data mining, and new initiatives are being formed. So, as time goes by, we will likely get good answers to many of the questions we have posed.

8.4.6 USER INTERFACE ASPECTS

As in any kind of system, having a good user interface is critical to mining. Note that some of the early database management systems had very primitive user interfaces. Therefore, users had to spend a great deal of time writing SQL (structured query language) queries and application programs. After much work, current database systems have excellent user interface tools. User interface tools are also being developed for multimedia databases. These include tools for generating queries, applications programs, and reports. Various multi-modal interfaces are also being provided for database management.

User interface support for current data mining systems is fairly primitive. As mentioned earlier, visualization tools are being developed to help with data mining, but tools for generating queries, application programs to carry out data mining, and reports are not sophisticated. To make data mining a success we need better user interface tools. Computer scientists and technologists are not the only ones who should be involved in developing such tools. Interactions between technologists, scientists, psychologists, and human computer specialists are necessary to develop better tools. Figure 8.31 illustrates an example user interface for data mining. The interface has buttons not only for generating data mining queries, applications for data mining, and reports, but also for selecting the outcomes desired, approaches to be followed, and techniques to be utilized.

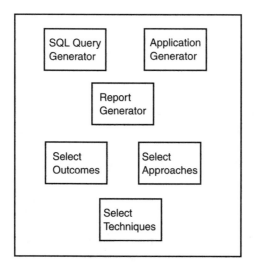

FIGURE 8.31 Example user interface for mining.

8.5 DATA MINING OUTCOMES, APPROACHES, AND TECHNIQUES

8.5.1 OVERVIEW

This section focuses on concepts in data mining. In particular, what possible outcomes can one expect, what are the approaches or methodologies used, and what are the data mining techniques used? Various data mining and machine learning text books, such as the ones by Mitchell,[85] Berry and Linoff,[17] and Adriaans and Zantinge,[18] have focused mainly on the topics discussed in this chapter. In particular, Berry and Linoff[17] have provided an excellent discussion of the outcomes, approaches and techniques for data mining.* Therefore, we will only discuss the essential points in this chapter. For further reading, the reader is referred to the references provided. Many of the techniques discussed here can be applied for multimedia data. For example, one can find associations in text as well as anomalies in images.

The outcomes of data mining are also referred to as data mining tasks or types. These are the results that one can expect to see as a consequence of data mining. We discuss some of the tasks in Appendix C. The data mining outcomes include classification, clustering, prediction, estimation, and affinity grouping. Note that there is no standard terminology. Therefore, different papers and texts have sometimes used different terms to mean the same thing.

---

* In fact, Berry and Linoff call these terms data mining tasks, methodologies, and techniques.[17] As we have mentioned, there is no standard terminology for data mining. We hope that the data mining community will eventually standardize various terms.

The approaches to data mining, also referred to as methodologies, are top-down, bottom-up, or hybrid. In addition, the methods can also be directed or undirected. Directed techniques are also sometimes called supervised learning, while undirected techniques are called unsupervised learning.

Data mining techniques are the algorithms employed to carry out data mining. There has been some confusion between techniques and outcomes. For example, a collection of data mining techniques is used for market basket analysis. These techniques are now known as market basket analysis techniques. However, market basket analysis is also an application dealing with determining which items are purchased together in a supermarket. Therefore, there is some confusion within the community as to what these terms mean. We expect progress to be made with respect to terminology as the technology matures and standards are developed.

In general, to carry out data mining for a specific application, we first have to decide on the type of outcome expected from the process. We then have to determine the techniques to be employed to achieve the expected outcome. Finally, we have to determine whether to steer the process in a top-down, bottom-up, or hybrid fashion. Figure 8.32 illustrates these steps.

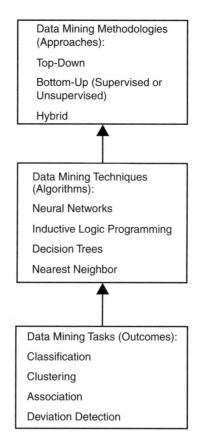

FIGURE 8.32 Data mining tasks, techniques, and methodologies.

This chapter is devoted to the outcomes, approaches, and techniques for data mining. The outcomes are discussed in Section 8.5.2. Approaches are the subject of Section 8.5.3. Data mining techniques are addressed in Section 8.5.4.

8.5.2 Outcomes of Data Mining

The outcomes of data mining are also referred to as data mining tasks or types. Some of them are discussed here. In a task called classification, the mining tool examines the features of a new entity, examines a predefined set of classes, and classifies the entity as belonging to a particular class if common features are extracted. For example, a class of mammals has attributes that describe a mammal. If a living entity has to be classified and satisfies the properties for a mammal, then it can be classified as a mammal. Classification is carried out by developing training sets with preclassified examples and then building a model that fits the description of the classes. This model is then applied to the data not yet classified and results are obtained. In summary, with classification, a group of entities is partitioned based on a predefined value of some attribute. For example, one can classify text and images depending on the content.

Estimation and prediction are two other data mining tasks. In the case of estimation, based on the spending patterns of a person and his age, one can estimate his salary or the number of children he has. Prediction tasks predict the future behavior of some value. For example, based on the education of a person, his current job, and the trends in the industry, one can predict that his salary will be a certain amount by the year 2005. As another example, depending on the patterns observed in newspaper articles, one can predict certain events in the future.

One task that is extremely useful is affinity grouping. This is also sometimes referred to as making associations and correlations. Essentially, this determines the items that go together. Who are the people that travel together? What are the items that are purchased together? While prediction is some future value and estimation is an estimated value, affinity grouping makes associations between current values.

Clustering is a data mining task that is often confused with classification. While classification classifies an entity based on some predefined values of attributes, clustering groups similar records that are not based on some predefined values. That is, when one classifies a group of people, he essentially has predefined classes based on values of some attributes. In the case of clustering, one does not have these predefined classes. Instead, clusters are formed by analyzing the data. For example, suppose we want to find something interesting about a group of people in a community. We do not have any predefined classes. We then analyze their spending patterns on automobiles. Clusters are formed based on the analysis: group X prefers Volvos, group Y prefers Saabs, and group Z prefers Mercedes. Once the clusters are obtained, each cluster can be examined and mined further for other outcomes such as estimation and classification.

Other data mining tasks include deviation analysis and anomaly detection. For example, John usually goes shopping after he goes to the bank, but last week he went to church after shopping. Anomaly detection is a form of deviation detection and is used for applications such as fraud detection and medical illness detection.

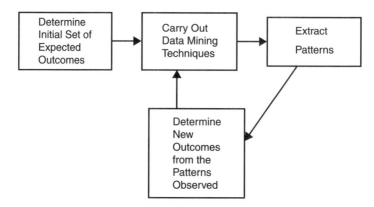

FIGURE 8.33 Data mining outcomes.

Some consider tasks like summarization and semantic content exploration, e.g., understanding the data, to be data mining tasks.

Here are some of our observations. While one can see differences between the different tasks, there are often similarities as well. A common question is, what is the difference between affinity grouping and estimation? There is still no theory behind data mining that is precise as to the notions and definitions. It is still rather ad hoc. Furthermore, while neural networks are good for clustering, one cannot make definite statements such as technique A is used for task X and technique B is good for task Y. It is largely trial and error, and one gets better with experience. Therefore, without doing a lot of data mining, it is difficult to determine what is best for a particular situation. Figure 8.33 illustrates the process of determining data mining outcomes.

8.5.3 APPROACHES TO MULTIMEDIA DATA MINING

Berry and Linoff[17] have clearly explained the approaches to data mining. They call these approaches methodologies for data mining. These methodologies are not the outcomes nor are they the techniques. They are the steps one would take to mine data. Once the outcomes are determined, how does one go about performing the mining? This section addresses this issue.

Essentially, there are two approaches: top-down and bottom-up. One can combine the two and have a hybrid approach. The top-down approach starts with some idea, pattern, or hypothesis. For example, a hypothesis could be, "All those who live in Concord, Massachusetts earn a minimum of $50,000." Then one would start querying the database and test his ideas and hypothesis. If he finds something that does not confirm his hypothesis, he has to revise his hypothesis. A lot of statistical reasoning is used for this purpose. In general, hypothesis testing is about generating ideas, developing models, and then evaluating the model to determine if the hypothesis is valid or not. Developing the model is a major challenge. If the model is not good, one cannot rely on the outcome. The models could simply be a collection of rules of the form, "If a person lives in New York, then he owns a house worth more than $300,000." To evaluate the model, one needs to query the database. In the above example, one could query to select all those living in New York in homes costing less than $300,000.

In the bottom-up approach to data mining, there is no hypothesis to test. This is much harder because the tool has to examine the data and then come up with patterns. The bottom-up approach can be directed or undirected. In directed data mining, also referred to as supervised learning in the machine learning literature, one must have some idea what he is looking for. For example, who often travels with John to New York? What item is often purchased with milk? Like the top-down approach, models are developed and evaluated based on the data analyzed. With undirected data mining, also called unsupervised learning in the machine learning literature, one has no idea what he is looking for. He asks the tool to find something interesting. For example, in image data mining, the data mining tool can go about finding something that it thinks is unusual. As before, a model is developed and evaluated with the data. Once something interesting is found, directed data mining can be conducted.

The hybrid approach is a combination of both top-down and bottom-up mining. For example, we can start with bottom-up mining, analyze the data, and then discover a pattern. This pattern could be a hypothesis, and we can do top-down mining to test the hypothesis. As a result, we can find new patterns that create a new hypothesis. That is, the tool can switch between top-down and bottom-up mining and again between directed and undirected mining. Data mining approaches are illustrated in Figure 8.34.

8.5.4 DATA MINING TECHNIQUES AND ALGORITHMS

The algorithms and techniques employed to perform the data mining are most important. Data mining techniques are numerous. These include statistical analysis, machine learning, and other reasoning techniques. Only a few of them are discussed here. Note that we have not distinguished between techniques and algorithms. One can argue that while techniques describe a broad class of procedures to carry out mining, algorithms go into more detail. For example, while link analysis can be regarded as a data mining technique, one can employ various algorithms, such as intelligent searching and graph traversal, to carry out link analysis. The outcome of link analysis would be associations between various entities.

One popular class of data mining techniques is called market basket analysis. This refers to techniques that group items together. For example, which items go together, who travels with whom together, and what events occur together? The actual techniques employed to do market basket analysis are intelligent searching and pruning that search. Many of the intelligent search techniques developed for artificial intelligence are being employed for market basket analysis. Searching the entire search space would be combinatorially explosive. Therefore, the challenge is to determine how to eliminate unnecessary items from the search space. Various papers and books have given examples of market basket analysis from supermarket purchases. The idea is to make a list of all purchases for a certain period and then analyze these purchases to see which items are often purchased together. There will be some obvious answers such as bread and milk or bread and cheese. What the decision maker is looking for is some of the non-obvious answers like bread and soy sauce.

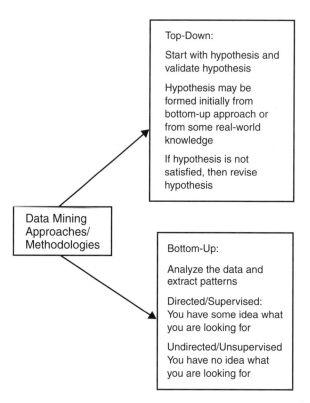

FIGURE 8.34 Some data mining approaches.

Another data mining technique is decision trees. This is a machine learning technique and is used extensively for classification. Records and objects are divided into groups based on some attribute value. For example, the population may be divided into two groups, one consisting of those who earn less than $100,000 and the other consisting of those who earn more than $100,000. Each of the groups is then divided further based on some value such as age. That is, each class is divided into subclasses depending on whether the members are over or under the age of 50. Each of the subclasses can then be further divided depending on marital status. Subsequently, a tree structure is formed with leaves at the end; the decision tree is then used for training. Then, as new data appears to be analyzed, the training examples are used to classify the data.

Neural networks are another popular data mining technique that has been around for while. A neural network is essentially a collection of input signals, nodes, and output signals. The network is first trained with training sets and examples. Once the learning is over, new patterns are given to the network. The network then uses its training experience and analyzes the new data. Data may be used for clustering, identifying entities, deviation analysis, and various other data mining tasks.

Inductive logic programming is a machine learning technique that is of special interest to these authors because of our own research. It originated from logic programming. Instead of deducing new data from existing data and rules, inductive

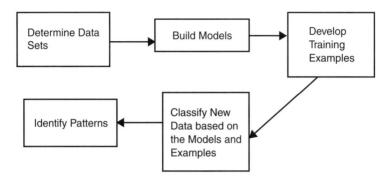

FIGURE 8.35 Operation of data mining techniques.

logic programming is all about inducing rules from analyzing data. It uses a variation of the resolution principle in theorem proving for discovering rules. Since inductive logic programming is of interest to us, we have devoted the next chapter to that topic.

Several other data mining techniques are used today, including link analysis techniques, which are a collection of techniques to find associations and relationships between records; automatic cluster detection techniques, which are a collection of techniques to find clusters; techniques to find association rules that are similar to link analysis; and nearest neighbor techniques, which are a collection of techniques that analyze new data based on its neighbors. For example, in the nearest neighbor techniques, if a situation has to be analyzed, the database is examined to see if there are neighbors with similar properties. Conclusions are then drawn about the new situation. These techniques employ distance functions to determine the closeness between data entities. It is assumed that points in space that are close together have similar properties. So, the points in space of new data are based on some predefined computation. These points of space are evaluated to find out how close they are to known data points, and then their properties are determined. Association rule-based techniques are popular among database researchers. These techniques essentially examine data in a database and come up with associations between the entities. In many ways, the techniques are similar to those used for link analysis.

Other data mining techniques include those based on genetic algorithms, fuzzy logic, rough sets, concept learning, and simple rule-based reasoning. A detailed discussion of these techniques is available in the literature.[17,79,85] As mentioned, there is some overlap between the different techniques that have been proposed. These techniques have been taken from statistics, data management, and machine learning. As we make progress toward integrating the various data mining technologies, we can expect more sophisticated techniques to be developed. Figure 8.35 illustrates the way data mining techniques operate.

8.6 SUMMARY

This chapter has taken various data mining prerequisites and examined their impact on multimedia data. We started with a discussion of multimedia data mining technologies, including multimedia database management, multimedia warehousing, as well some traditional data mining technologies such as statistical reasoning and visualization. We also provided an overview of architectural support. We then discussed the need for data mining and gave several examples. Finally, an overview of data mining techniques was described. Many of these techniques may also be applied for multimedia data mining.

Now that we have some background on multimedia data mining, including technologies and techniques for multimedia data mining, we are now ready to discuss issues concerning mining text, images, video, and audio data. That topic will be addressed in the next chapter.

9 Mining Text, Image, Video, and Audio Data

9.1 OVERVIEW

Now that we have provided an overview of multimedia databases and multimedia data mining, and discussed some of the essential concepts in terms of architectures, data models, and functions, this chapter discusses the issues involved in mining and extracting information from text, images, video, and audio data.

As stated earlier, multimedia data includes text, images, video, and audio. Text and images are still media, while audio and video are continuous media. The issues surrounding still and continuous media are somewhat different and have been explained in various texts and papers, for example, Prabhakaran.[102] This section considers text, image, video, and audio data and how such data can be mined. What are the differences between mining multimedia data and topics such as text, image, and video retrieval? What is meant by mining such data? What are the developments and challenges?

Data mining has an impact on the functions of multimedia database systems discussed in the previous chapter. For example, query processing strategies have to be adapted to handle mining queries if a tight integration approach between the data miner and the database system is taken. This will then have an impact on storage strategies. Furthermore, the data model will also have an impact on storage strategies. At present, many mining tools work on relational databases. However, if object-relational databases are to be used for multimedia modeling, then data mining tools have to be developed to handle such databases.

Sections 9.2, 9.3, 9.4, and 9.5 will discuss text, image, video, and audio mining, respectively. Section 9.6 briefly discusses the issues of mining combinations of data types. The chapter is summarized in Section 9.7.

9.2 TEXT MINING

9.2.1 OVERVIEW

The previous overview of multimedia data mining and the techniques for data mining allows us to focus now on mining text, images, audio, and video data. This section provides an overview of text mining.

Unlike structured data, text is a difficult data type because it is often challenging to figure out what is really meant by the text. For example, by changing one word, an entire idea may be changed, such as, "I think John is clever" versus "I believe that John is clever." Also, text retrieval in general, and information retrieval in particular, has been around for many years. Document management systems evolved into text processing or information retrieval systems. What is the difference between text mining and text processing? We believe that the situation is similar to the difference between database management and data mining. Therefore, this section discusses text processing/retrieval and text mining and then provides a taxonomy for text mining.

The organization of this section is as follows. Section 9.2.2 discusses text retrieval/processing. Text mining is discussed in Section 9.2.3. Finally, taxonomy for text mining is provided in Section 9.2.4.

9.2.2 TEXT RETRIEVAL

A text retrieval system is essentially a database management system for handling text data. Text data may be documents such as books, journals, magazines, etc. There are various issues that need to be considered. Some of them are discussed in this section.

One needs a good data model for document representation. A considerable amount of work has gone into developing semantic data and object models for document management. For example, a document has paragraphs and a paragraph has sections, as shown in Figure 9.1. One also needs an architecture for the system. A functional architecture would have the following modules: query manager, browser, editor, update manager (which may overlap with the editor), storage manager, and metadata manager. In addition, an integrity and security manager is necessary to maintain integrity and security. An architecture for a text processing system is illustrated in Figure 9.2.

Querying documents can be based on many factors. One can specify keywords and request the retrieval of any documents containing those keywords. One can also retrieve documents that have some relationship with one another. The questions is,

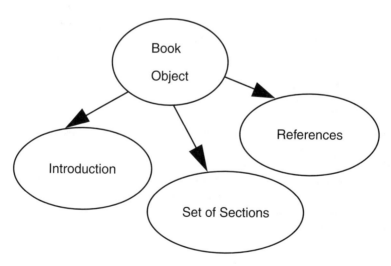

FIGURE 9.1 Data model for text.

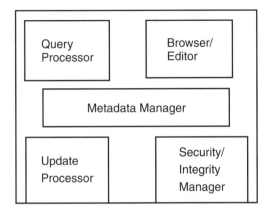

FIGURE 9.2 Functional Architecture for a text processing system.

however, is a system that supports relationships a text processing or text mining system? The answer is very subjective. Our discussion on taxonomy includes different levels of complexity for document management. The following section deals with text mining.

9.2.3 TEXT MINING

So much information is now in textual form, such as data on the Web, library data, or electronic books, among others. One of the problems with text data is that it is not structured as relational data. In many cases, it is completely unstructured and in other cases it is only semistructured. Semistructured data, for example, is an article that has a title, author, abstract, and paragraphs. The paragraphs are not structured, however, the format is.[23]

Information retrieval systems and text processing systems have been developed for more than a few decades. Some of these systems are quite sophisticated and can retrieve documents using specific attributes or keywords. There are also text processing systems that can retrieve associations between documents. We are often asked what the difference is between information retrieval systems and text mining systems?

We define text mining as data mining on text data. Text mining is all about extracting patterns and associations previously unknown from large text databases. The difference between text mining and information retrieval is analogous to the difference between data mining and database management, i.e., there is really no clear difference. Some recent information retrieval and text processing systems do discover associations between words and paragraphs and, therefore, can be regarded as text mining systems.

Next, let us examine the approaches to text mining. Note that many of the current tools and techniques for data mining work are for relational databases. Rarely do we hear about data mining tools for data in object-oriented databases. Therefore, current data mining tools cannot be directly applied to text data. Some of the current directions in mining unstructured data include the following:

- Extract data and/or metadata from the unstructured databases possibly by using tagging techniques, store the extracted data in structured databases, and apply data mining tools to the structured databases. This is illustrated in Figure 9.3.
- Integrate data mining techniques with information retrieval tools so that appropriate data mining tools can be developed for unstructured databases. This is illustrated in Figure 9.4.
- Develop data mining tools to operate directly on unstructured databases. This is illustrated in Figure 9.5.

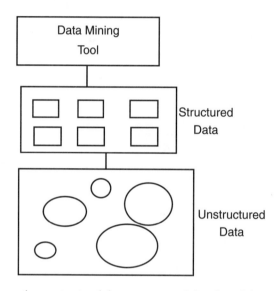

FIGURE 9.3 Converting unstructured data to structured data for mining.

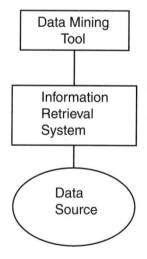

FIGURE 9.4 Augmenting an information retrieval system.

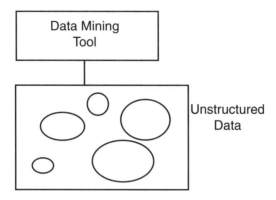

FIGURE 9.5 Mining directly on unstructured data.

While converting text data into relational databases, one has to be careful that there is no loss of key information. As stated earlier, without good data, data mining will not be effective or produce useful results. One needs to create a sort of warehouse before mining the converted database. This warehouse is essentially a relational database that has the essential data from the text. In other words, one needs a transformer that takes a text corpus as input and outputs tables that contain, for example, the keywords from the text.

For example, in a text database that has several journal articles, one could create a warehouse with tables containing the following attributes: author, date, publisher, title, and keywords. Associations can be formed from the keywords. The keywords in one article could be "Belgium, nuclear weapons" and the keywords in another article could be "Spain, nuclear weapons." The data miner might make the association that authors from Belgium and Spain write articles on nuclear weapons.

Note that we are only in the beginning stages of developing text mining. In the longer term, we want to develop tools directly to mine text data. These tools will have to read and understand the text, put out pertinent information about the text, and then make associations between different documents. We are far from developing such sophisticated text mining tools. However, the work reported by Tsur et al.[131] is the first step in the right direction toward more evolved text mining. Some interesting early work on text mining was reported by Feldman and Dagan.[43]

9.2.4 Taxonomy for Text Mining

Now that we have discussed text retrieval and mining, we can provide a taxonomy of text mining. This taxonomy is illustrated in Figure 9.6. At the bottom is text retrieval, and at the top is true text mining systems. In between there may be multiple levels. At one level, there may be an intelligent text mining system that can make deductions and inferences. At another layer, there may be a text understanding system. At a third layer, there may be a semi-text mining system that can make partial correlations. At a fourth layer, there may be a system that extracts structured data from text and mines the structured data. As we go higher and higher, we reach a true text mining system where text is mined directly.

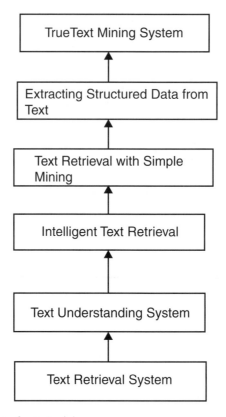

FIGURE 9.6 Taxonomy for text mining.

Note that this text mining taxonomy is somewhat subjective and depends on whose viewpoint it is. Nevertheless, there is a distinction between text mining and text retrieval. At one end is a simple query system for retrieving documents. At the other end is a true text mining system. In between are many layers in increasing levels of sophistication.

9.3 IMAGE MINING

9.3.1 OVERVIEW

Unlike structured data, image mining can be difficult because of the difficulties involved in analyzing an image. In fact, images are harder to understand than text, and, as a result, they are harder to mine. For example, changing one part of an image can change its whole meaning, such as changing an image from a ship to a submarine. Also, image processing in general, and image retrieval in particular, has been studied for several years now. Image retrieval systems evolved into image processing systems. What is the difference, then, between image mining and image processing? We believe that the discrepancy is similar to the difference between data mining and database systems. This is also similar to the difference between text mining and text retrieval. Therefore, this section discusses image processing/retrieval and image mining and then provides a taxonomy for image mining.

The organization of this section is as follows. Section 9.3.2 discusses image retrieval/processing. Image mining is discussed in Section 9.3.3. Finally, a taxonomy for image mining will be given in Section 9.3.4.

9.3.2 IMAGE RETRIEVAL

An image retrieval system is essentially a database management system for handling image data. Image data can be x-rays, pictures, satellite images, and photographs. There are various issues that need to be considered; we discuss some of them in this section.

A good data model is necessary for image representation. Some work has gone into developing semantic data and object models for image management.[120] For example, an image may consist of a right image and a left, as shown in Figure 9.7 (one example is an x-ray of the lungs). An architecture is then needed for the system. A functional architecture would have the following modules: query manager, browser, editor, update manager (which may overlap with the editor), storage manager, and metadata manager. In addition, an integrity and security manager is needed to maintain integrity and security. An architecture for an image processing system is illustrated in Figure 9.8.

Querying images can be based on many factors. One can extract text from images and then query the text. One can tag images and then query the tags. One can also retrieve images from patterns. For example, an image may contain several squares. With a picture of a square, one could query the image and retrieve all the squares in the image. Is a system that finds associations or relationships an image processing system or an image mining system? The answer is very subjective. Our discussion of taxonomy includes different levels of complexity for image management. The next section discusses image mining.

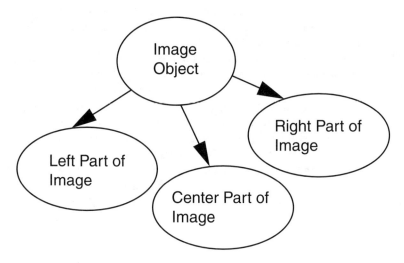

FIGURE 9.7 Data model for an image.

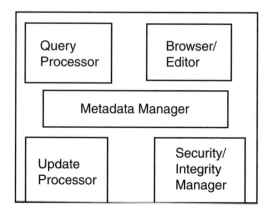

FIGURE 9.8 Functional architecture for an image processing system.

9.3.3 IMAGE MINING

If text mining is still in the early research stages, image mining is an even more immature technology. This section examines this area and discusses the current status and challenges.

Image processing has been around for quite a while. Image processing applications exist in various domains, including medical imaging for cancer detection, processing satellite images for space and intelligence applications, and handling hyperspectral images. Images include maps, geological structures, biological structures, and many other entities. Image processing has dealt with areas such as detecting patterns that deviate from the norm, retrieving images by content, and pattern matching.

What is image mining? How does it differ from image processing? Again, there are no clear cut answers. While image processing focuses on detecting abnormal patterns and retrieving images, image mining is all about finding unusual patterns. Therefore, one could say that image mining deals with making associations between different images from large image databases.

Clifton[24] has undertaken work in image mining. Initially, his plan was to extract metadata from images and then carry out mining on the metadata. This would essentially mean mining the metadata in relational databases. However, after some consideration, the consensus was that images could be mined directly. The challenge, then, is to determine what type of mining outcome is most suitable. One can mine for associations between images, cluster images, classify images, as well as detect unusual patterns. One area of research being pursued by Clifton and associates is to mine images and find out whether there is anything unusual. The approach is to develop templates that generate several rules about the images and, from there, apply the data mining tools to see if unusual patterns can be obtained. However, the mining tools do not tell us why these patterns are unusual. Figure 9.9 shows an image with some unusual patterns.

Note that detecting unusual patterns is not the only outcome of image mining. However, image mining is still in its early stages. More research on image mining

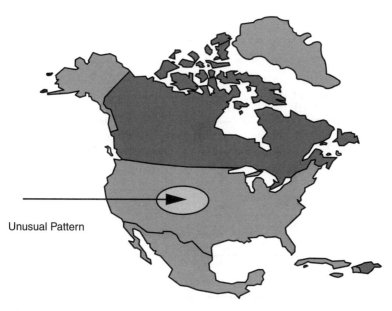

Unusual Pattern

FIGURE 9.9 Image mining.

needs to be conducted to see if data mining techniques could be used to classify, cluster, and associate images. Image mining is an area with applications in many domains including space images, medical images, and geological images.[91]

9.3.4 TAXONOMY FOR IMAGE MINING

Now that we have discussed image retrieval and mining, we can provide a taxonomy of image mining. This taxonomy is illustrated in Figure 9.10. At the bottom is image retrieval and at the top is true image mining systems. In between there may be multiple levels. At one level there may be an intelligent image mining system that can make deductions and inferences. At another layer, there may be a semi-image mining system that can make partial correlations. At a third layer, there may be a system that extracts structured data from images and mines the structured data. As we go higher and higher, we reach a true image mining system.

Note that this image mining taxonomy is somewhat subjective and depends on viewpoint. Nevertheless, there is a distinction between image mining and image retrieval. At one end is a simple query system for retrieving images, and at the other end is a true image mining system. In between are many layers in increasing levels of sophistication.

9.4 VIDEO MINING

9.4.1 OVERVIEW

Let us now examine video mining. Unlike structured data, video data is hard to understand. Video is even harder to understand than text and images because video

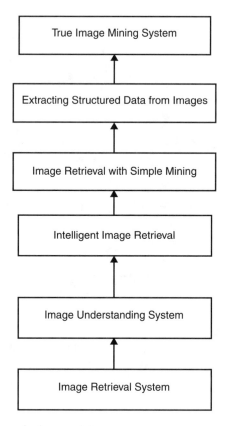

FIGURE 9.10 Taxonomy for image mining.

is continuous media. In fact, video consists of several moving images. For example, by changing one video script, the whole meaning may be changed. Unlike text retrieval and image processing, only recently have video retrieval systems been developed. What is the difference between video mining and video retrieval/processing? We believe that the difference is similar to the difference between data mining and database management. We can also draw an analogy to the difference between text mining and text retrieval as well as the difference between image mining and image retrieval. Therefore, this section discusses video processing/retrieval and video mining and then provides a taxonomy for video mining.

The organization of this section is as follows. Section 9.4.2 discusses video retrieval/processing. Video mining is discussed in Section 9.4.3. Finally, a taxonomy for video mining is provided in Section 9.4.4.

9.4.2 VIDEO RETRIEVAL

A video retrieval system is essentially a database management system for handling video data. Various issues need to be considered, and some of them are discussed in this section.

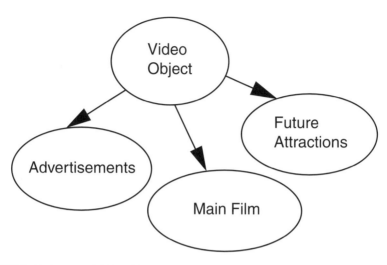

FIGURE 9.11 Data model for video.

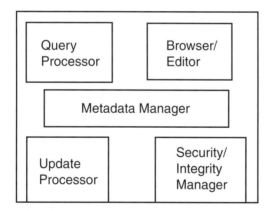

FIGURE 9.12 Functional architecture for a video processing system.

One needs a good data model for video representation. Some work has gone into developing semantic data and object models for video data management.[139] For example, a video object could have advertisements, main film and coming attractions as shown in Figure 9.11. An architecture is then needed for the system. A functional architecture would have the following modules: query manager, browser, editor, update manager (which may overlap with the editor), storage manager, and metadata manager. In addition, an integrity and security manager is needed to maintain integrity and security. An architecture for a video processing system is illustrated in Figure 9.12.

Querying documents can be based on many factors. One can extract text from the video and query the text. One can also extract images from the video and query the images. One can store short video scripts and carry out pattern matching, e.g.,

Video Frames

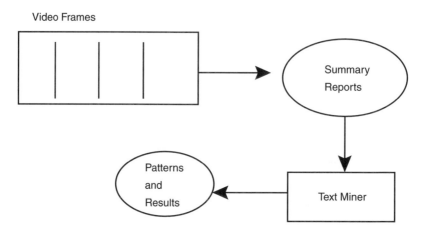

FIGURE 9.13 Mining text extracted from video.

"find the video that contains the following script." One can also video documents that have some relationship to one another. Is a system that finds relationships a video processing system or a video mining system? The answer is very subjective. Our discussion of taxonomy includes different levels of complexity for video data management. The next section discusses video mining.

9.4.3 VIDEO MINING

Mining video data is even more complicated than mining image data (Figure 9.13). One can regard video as a collection of moving images, much like animation. Video data management has been the subject of much research. The important areas include developing query and retrieval techniques for video databases including video indexing, query languages, and optimization strategies. It is important again to clarify the difference between video information retrieval and video mining. Unlike image and text mining, there is no clear idea of what is meant by video mining. For example, one can examine video clips and find associations between different clips. Another example would be to find unusual patterns in video clips. How is this different, however, from finding unusual patterns in images? The first step to successful video mining is to have a good handle on image mining.

Let us examine pattern matching in video databases. Should one have predefined images and then match these images with the video data? Is there any way to do pattern recognition in video data by specifying what one is looking for and then performing feature extraction for the video data? If this is video information retrieval, what is video data mining? To be consistent with our terminology, we can say that finding previously unknown correlations and patterns in large video databases is video mining. So, by analyzing a video clip or multiple video clips, one draws conclusions about some unusual behavior. People in a video who are unlikely to be there yet have appeared two or three times may mean something significant. Another approach to the problem is to capture the text in video format and try to make the associations one would carry out with text using video data instead.

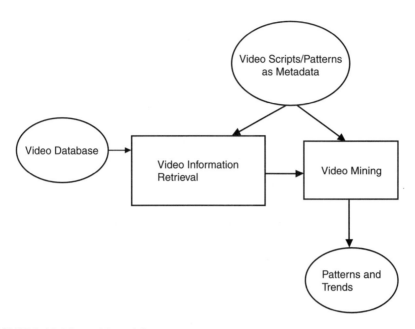

FIGURE 9.14 Direct video mining.

Unlike text and image mining where our ideas have been somewhat less vague, our discussion of video mining is quite preliminary. This is mainly because there is so little known about video mining. Even the word video mining is something very new, and, to date, we do not have any concrete reported results. We do have a lot of information on analyzing video data and producing summaries. One could mine these summaries, which would amount to mining text, as shown in Figure 9.13. One good example of this effort is the work of Merlino et al.[83] concerning summarizing video news. Converting the video mining problem to a text mining problem is reasonably well understood. However, the challenge is mining video data directly, and, more importantly, knowing what to mine. With the emergence of the Web, video mining becomes even more important. One example of direct video mining is illustrated in Figure 9.14.

Another noteworthy point is that one can use techniques for image mining to mine video data. That is, one can detect abnormal patterns in video data. One can use neural networks to train normal video, for example, who usually appears with whom, and then if something abnormal occurs, such as "John appearing with Mary" for the first time, then the system will flag this as something abnormal. Anomaly detection for video mining needs further investigation.

9.4.4 TAXONOMY FOR VIDEO MINING

Now that we have discussed video retrieval and video mining, we can provide a taxonomy of video mining. This taxonomy is illustrated in Figure 9.15. At the bottom is video retrieval and at the top is true video mining systems. In between there may be multiple levels. At one level there may be an intelligent video mining system that

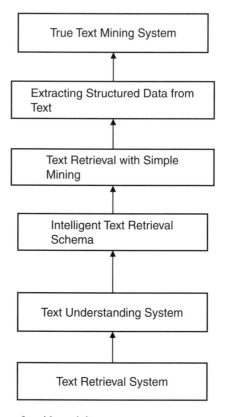

FIGURE 9.15 Taxonomy for video mining.

can make deductions and inferences. At another layer, there may be a semi-video mining system that can make partial correlations. At a third layer, there may be a system that extracts structured data from video and mines the structured data. As we go higher and higher, we reach a true video mining system.

Note that this video mining taxonomy is somewhat subjective and depends on viewpoint. Nevertheless, there is a distinction between video mining and video retrieval. At one end is a simple query system for retrieving video, and at the other end is a true video mining system. In between are many layers in increasing levels of sophistication.

9.5 AUDIO MINING

9.5.1 OVERVIEW

Next, let us examine audio mining. Unlike structured data, audio data is hard to understand. Like video, audio is even harder to understand than text and images because audio is continuous media. By changing the tone of the audio, the whole meaning may be changed. Unlike text retrieval and image processing and even video retrieval, hardly any work has been carried out on audio retrieval, although recently there has been a lot of research on speech processing systems. What is the difference between audio mining and audio retrieval/processing? We believe that the situation is similar to the difference between data mining and database management. We can also draw an analogy to the difference between text mining and text retrieval, image mining and image retrieval, and video mining and video retrieval. Therefore, this section discusses audio processing/retrieval and audio mining and then provides a taxonomy for audio mining.

The organization of this section is as follows. Section 9.5.2 discusses audio retrieval/processing. Audio mining is discussed in Section 9.5.3. Finally, a taxonomy for audio mining is provided in Section 9.5.4.

9.5.2 AUDIO RETRIEVAL

An audio retrieval system is essentially a database management system for handling audio data. Various issues need to be considered; some of them are discussed in this section.

A good data model is necessary for audio representation. Some work has gone into developing semantic data and object models for audio data management.[139] For example, an audio object can have speech, applause, and questions, as shown in Figure 9.16. Then you need an architecture for the system. A functional architecture would have the following modules: query manager, browser, editor, update manager (which may overlap with the editor), storage manager, and metadata manager. In addition, an integrity and security manager is necessary to maintain integrity and security. An architecture for an audio processing system is illustrated in Figure 9.17.

Querying documents can be based on many factors. One might extract text from the audio and query the text. One can store short audio scripts and carry out pattern matching, e.g., "find the audio that contains the following script." One could also record documents that have some relationship to one another. Is a system that finds

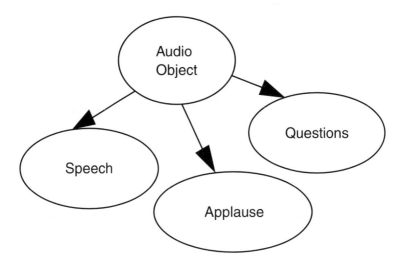

FIGURE 9.16 Data model for audio.

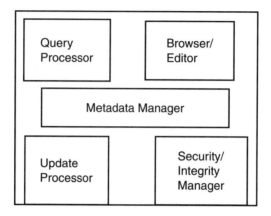

FIGURE 9.17 Functional architecture for an audio processing system.

relationships an audio processing system or an audio mining system? The answer is very subjective. Our discussion of taxonomy includes different levels of complexity for audio data management. Audio mining is discussed in the next session.

9.5.3 AUDIO MINING

Since audio is a continuous media type like video, the techniques for audio information processing and mining are similar to video information retrieval and mining. Audio data can be in the form of radio, speech, or spoken language. Even television news has audio data, and, in that case, audio may be integrated with video and possibly text to capture the annotations and captions.

To mine audio data, one can convert it into text using speech transcription and other techniques, such as keyword extraction, and then mine the text data, as illustrated in Figure 9.18. On the other hand, audio data can also be mined directly by

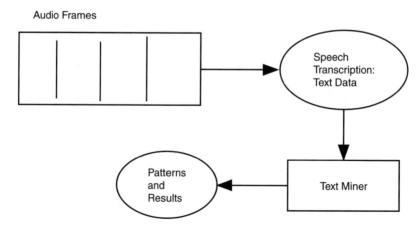

FIGURE 9.18 Mining text extracted from audio.

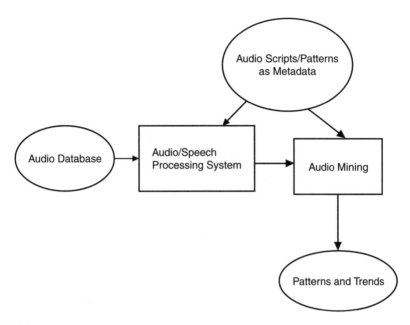

FIGURE 9.19 Direct audio mining.

using audio information processing techniques and then mining selected audio data. This is illustrated in Figure 9.19.

In general, audio mining is even more primitive than video mining. While only a few papers have appeared on text mining and fewer still on image and video mining, work on audio mining is just beginning.

Note that one can use image mining techniques for audio mining, too. That is, one can detect abnormal patterns in audio data. Neural networks can train normal audio data, such as who speaks what and when something abnormal occurs for the first time, and then the system will flag this as something abnormal. Anomaly detection for audio mining needs further investigation.

9.5.4 TAXONOMY FOR AUDIO MINING

Now that we have discussed audio retrieval and mining, we can provide a taxonomy of audio mining. This taxonomy is illustrated in Figure 9.20. At the bottom is audio retrieval, and at the top is true audio mining systems. In between there may be multiple levels. At one level there may be an intelligent audio mining system that can make deductions and inferences. At another there may be a semi-audio mining system that can make partial correlations. At a third layer, there may be a system that extracts structured data from audio and mines the structured data. As we go higher and higher, we reach a true audio mining system.

Note that this audio mining taxonomy is somewhat subjective and depends on viewpoint. Nevertheless, there is a distinction between audio mining and audio retrieval. At one end is a simple query system for retrieving audio. At the other end is a true audio mining system. In between are many layers in increasing levels of sophistication.

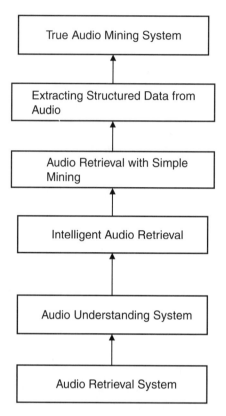

FIGURE 9.20 Taxonomy for audio mining.

9.6 MINING COMBINATIONS OF DATA TYPES

The previous sections discussed mining of individual data types such as text, images, video, and audio. Mining multimedia data requires mining combinations of two or more data types, such as text and images, text and video, or text, audio, and video. This section briefly discusses some of the issues of multimedia data mining.

Handling combinations of data types is very much like dealing with heterogeneous databases. For example, each database in the heterogeneous environment can contain data belonging to multiple data types. These heterogeneous databases could be integrated and then mined, or one can apply mining tools to the individual databases and then combine the results of the various data miners. These two scenarios are illustrated in Figures 9.21 and 9.22, respectively. In both cases, the DMP (distributed multimedia processor) discussed in Chapter 3 plays a role. If the data is to be integrated before being mined, this integration is carried out via the DMP. If the data is to be mined first, the data miner augments the corresponding MM-DBMS, and the results of the data miners are integrated via the DMP.

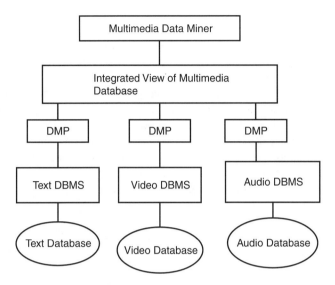

FIGURE 9.21 Integration and then mining.

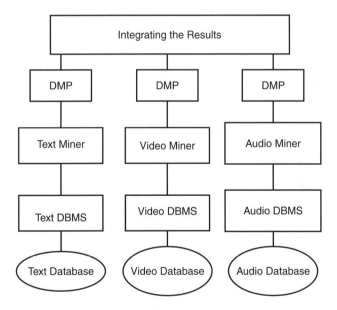

FIGURE 9.22 Integrating the results of various data miners.

Since there is much to be done regarding mining individual data types such as text, images, video, and audio, mining combinations of data types is still a challenge. Once we have a better handle on mining individual data types, we can then focus on mining combinations of data types.

9.7 SUMMARY

This chapter primarily addressed mining individual data types. Mining multimedia data would involve addressing a combination of two or more media types. As we learn more about mining text, images, video, and audio, we can expect progress to be made in the area of multimedia data mining. Mining combinations of data such as video and text, video, audio, and text, or image and text remains a challenge.

We believe that as progress is made in the areas of multimedia data management and data mining, we will begin to see tools emerge for mining multimedia data. At present, data mining tools work mainly on relational databases. However, in the future, we can expect to see multimedia data mining tools as well as tools for mining object databases. More research remains to be done in this area.

Conclusion to Part II

Part II, consisting of two chapters, described various aspects of multimedia data mining. Chapter 8 discussed multimedia data mining technologies, including multimedia data management and warehousing, statistical reasoning, and machine learning. We then discussed the need for multimedia data mining and gave examples. We also provided an overview of multimedia data mining techniques. Chapter 9 discussed issues of mining text, image, video, and audio data. We discussed the differences between text mining and text processing; image mining and image processing; video mining and video processing; and audio mining and audio processing. We described various taxonomies and also discussed issues of mining combinations of data types.

The technologies discussed in Parts I and II are necessary for various Web applications. That is, the large amounts of multimedia data on the Web have to managed and mined. Therefore, effective multimedia data management and mining techniques for the Web are needed. For applications such as e-commerce and e-business, it is necessary to manage not only text but also other types of data such as images, video, and audio. Part III addresses the important topic of managing and mining multimedia databases on the Web.

Part III

Multimedia for the Electronic Enterprise

INTRODUCTION

Part III, consisting of four chapters, describes multimedia data management and mining for the electronic enterprise. Chapter 10 provides an overview of multimedia for the Web and then discusses multimedia data management and multimedia data mining for Web applications such as e-commerce.

Chapter 11 describes multimedia for collaboration, knowledge management, and training. Note that collaboration, knowledge management, and training are important applications for the Web. We hear of collaborative commerce (c-commerce), where organizations collaboratively carry out e-commerce. Multimedia data management and mining are necessary for c-commerce, knowledge management and Web-based training.

Chapter 12 provides an overview of security and privacy issues. The World Wide Web has made it easier to exploit the privacy of individuals. We discuss security and privacy issues in general and then discuss the impact of managing multimedia data. Essentially, Web security issues as well as how they are applicable to e-commerce will be discussed.

Finally, Chapter 13 discusses standards, prototypes, and products for multimedia data management and mining. We begin with a discussion of standards such as XML (extensible markup language) and then provide an overview of multimedia database management prototypes and products. Finally, we discuss multimedia data mining prototypes and products.

10 Multimedia for the Web and E-Commerce

10.1 OVERVIEW

There are news organizations all over the world, from the British Broadcasting Corporation to the American Broadcasting Corporation to the Financial Times of London, the Wall Street Journal, and the Cable News Network. Many news organizations that are producers of multimedia information now want to put their information on the Web. There is so much data, however, that it is almost impossible to get quality presentations of multimedia data on the Web. This is especially true with continuous media such as video and audio.

There are network communication problems that have to be overcome. While there is progress and hardware is becoming less expensive, developing good software to ensure quality of service and timely access and presentation of data remains the challenge. There has been great progress in the past few years in implementing delayed broadcast services. For example, important speeches by heads of countries are posted on the Internet within minutes. But we still are a long way from high quality live video broadcast or live movies on the Web. That does not mean live entertainment is not yet possible. It is actually already available. However, the service we get today is not of good quality.

This chapter provides an overview of multimedia data management and mining for the Web and e-commerce. section 10.2 discusses some general considerations for multimedia data processing on the Web. Multimedia data management for the Web is the subject of section 10.3. Multimedia data mining for the Web and e-commerce is the subject of Section 10.4. We further elaborate on the discussions on multimedia data mining for the electronic enterprise in Sections 10.5, 10.6, and 10.7. Section 10.5 describes agents for multimedia data mining. Mining distributed multimedia databases is the subject of Section 10.6. Metadata for mining will be discussed in Section 10.7. The chapter is summarized in Section 10.8. For an overview of e-commerce, refer to Appendix F.

10.2 MULTIMEDIA DATA PROCESSING FOR THE WEB AND E-COMMERCE

The biggest consumers of multimedia data on the Web are the entertainment, broadcasting, and journalism industries. There is a huge market for multimedia Web data, and these organizations have tapped into only a small portion of it. As technology becomes more mature, we can expect major players in the entertainment industry to be very active in this field. In a way, this is all part of e-commerce. Although

143

multimedia for the Web does not deal with buying and selling music and videos on the Internet, it does deal with playing videos and music on the Internet. Some of the technical challenges for data management include synchronizing presentation with storage, security, and ensuring that quality of service is maintained. Another application area for multimedia on the Web is training and distance learning. That application is discussed in the next chapter.

Multimedia technology has numerous applications in e-commerce. Much of the data on the Web are text, images, video, and audio. This data has to be managed effectively. Sections 10.3 and 10.4 discuss multimedia data management and mining for the Web and e-commerce. Multimedia technology is also critical for training on the Web, for applications such as e-learning and e-education, among others. Multimedia is also an important aspect of e-entertainment. While e-learning and e-entertainment are part of e-business and not strictly e-commerce, one can use e-learning and e-entertainment for e-commerce purposes. That is, one could listen to parts of a video or audiotapes and then place orders via the Web. Developing multimedia presentations on the Web is also an important aspect. That is, users can gather different media information and put it together into a briefing for presentation. Multimedia for e-commerce is illustrated in Figure 10.1.

Multimedia technology also has another important application for e-commerce, and that is voice portals. Instead of filling out forms in text, one merely places a call via mobile phone or personal digital assistant. This facility is now available in a rather primitive form, however, we expect this to be a dominant technology in the future.

One of the major issues for multimedia data processing on the Web is data synchronization. That is, different media types have to be synchronized, and, for

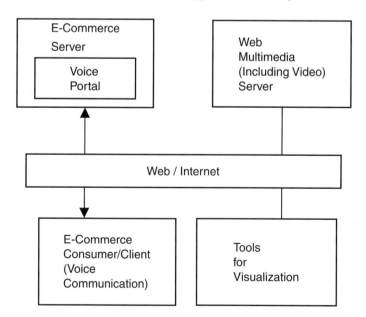

FIGURE 10.1 Multimedia/visualization for e-commerce.

live multimedia on the Web, timing constraints must be met. Multimedia data processing, quality of service, and real-time processing considerations are addressed in various parts of Part III.

One of the major developments over the past year is music on the Web, in particular, the developments with the company Napster (www.napster.com) and all the legal implications. The idea is to develop a distributed multimedia file system so that when a user requests a record, instead of going to a central server that stores the records, the system will approach any of the users who have already downloaded the record and then download that record to the machine of the requesting user. The advantage of this method is that the data are distributed, which avoids the bottleneck of a central server. However, there are many legal issues involved, for example, copyright of the record. There are no resolutions yet. The concept is good, however, and one can envisage video distribution on the Web based on the same idea.

Both live video and video on demand on the Web are receiving a lot of attention. The technology is almost in place, but we still have a long way to go before we get high quality timely multimedia audio and video on the Web. We believe that there are exciting challenges and opportunities in this area. Search engines also play a key role; many of them search for text, images, and video. One of the promising search engines for multimedia data is Google (www.google.com).

10.3 MULTIMEDIA DATA MANAGEMENT FOR THE WEB AND E-COMMERCE

This section discusses multimedia data management for the Web and e-commerce. Since much of the data on the Web is in the form of text, images, video, and audio, much of the discussion of Web database management and its applications to e-commerce found in Thuraisingham[128] is relevant to this section.

The information to be managed by the e-commerce site may be in large quantities, and, therefore, efficient multimedia data management techniques are necessary. Consider the multimedia database system to be part of the e-commerce server. Database system functions of interest include query processing, transaction processing, security management, and integrity management. In addition, metadata has to be extracted and maintained. Also, efficient storage management techniques are necessary. Figure 10.2 illustrates multimedia database systems and e-commerce.

Thuraisingham[128] discussed some of the metadata management issues for Web database management. Maintaining appropriate metadata is critical for intelligent browsing. As one goes through cyberspace, the metadata, which describes the navigation patterns, should be updated. This metadata is consulted periodically so that a user has some idea of where he is. Metadata, then, becomes like a map. Furthermore, the Internet metadata manager should continually give advice to the users.

Appropriate techniques are needed to manage metadata. These include querying and updating the metadata. The Internet environment is very dynamic, which means that the metadata must be updated continually as users browse through the Internet as well as when the data sources get updated. Furthermore, as new data sources get added, the changes have to be reflected in the metadata. Metadata may also include various security policies. The metadata must also be available to users in a timely

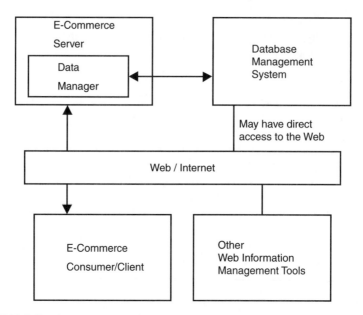

FIGURE 10.2 Database systems and e-commerce.

manner. Finally, appropriate models for metadata are also needed. These models may be based on the various data models or may utilize the models for text and multimedia data.

Metadata repositories may be included with the various data servers, or there may be separate repositories for the metadata. A scenario with multiple data servers and metadata repositories is illustrated in Figure 10.3. There is a lot of research being carried out on metadata management for the Internet.[11,12,84] However, a lot

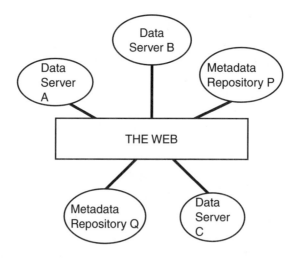

FIGURE 10.3 Metadata repositories on the Internet.

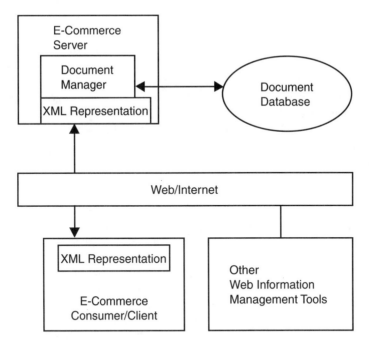

FIGURE 10.4 XML and e-commerce.

remains to be done before efficient techniques are developed for metadata representation and management. Defining the metadata is also a major issue.

Metadata is also a critical technology for e-commerce; it includes information about the e-commerce site, such as the internal information about how the site is structured. Metadata also includes information about the users of the site. Various ontologies are being developed for e-commerce. These ontologies are common definitions for various e-commerce terms. Finally, extensible markup language (XML) is being examined for e-commerce. XML specification for e-commerce would include information about specifying e-commerce site-specific data, information about the process for e-commerce, and domain-specific data such as securities information for financial transactions. With XML, both the clients and servers specify documents with common notations, i.e., domain-type definitions. Figure 10.4 illustrates an example of XML for carrying out Web transactions.

10.4 MULTIMEDIA DATA MINING FOR THE WEB AND E-COMMERCE

The previous section described multimedia data management for e-commerce. This section describes multimedia data mining for e-commerce. First, let us examine Web mining. Web mining is about making the data on the Web more manageable for users by creating associations and patterns. Web mining also involves mining the usage patterns to help users (i.e., consumers) carry out e-commerce and to support the e-commerce sites, brokers, and merchants.[29,47,58,87,138]

Corporations want a competitive edge and are exploring numerous ways to market effectively. Major corporations including retail stores have e-commerce sites, and customers can now order products from books to clothing to toys on these sites. One goal of e-commerce sites is to provide customized marketing. For example, user group A may prefer literature novels, whereas user group B may prefer mystery novels. Therefore, new literature novels have to be marketed to group A and new mystery novels have to be marketed to group B. An e-commerce site knows of these preferences by conducting data mining; the usage patterns have to be mined. In addition, a company may mine various public and private databases to obtain additional information about these users. For this reason, we included the data miner as part of the functional modules for an e-commerce site. An alternative is to embed the Web mining tool into the agent for the e-commerce site. This agent can be called the e-commerce server agent.

Web mining also helps the consumer by giving information about various e-commerce sites, what the best buys are, and how consumers can bargain with the seller. Therefore, Web miners are also part of the agents for the consumer. One can envision a scenario where Web mining tools are embedded into consumer and e-commerce agents. Web miners can also be directly embedded into consumer and e-commerce servers. Web mining may also be part of the brokers and mediators who do not act on behalf of anyone but carry out the functions of an honest broker. Note that we have differentiated between mediators and brokers: brokers are responsible for the brokering between consumers and e-commerce servers or between businesses in the case of business-to-business e-commerce, while mediators mediate between two parties to resolve heterogeneity and other differences that must be resolved before any brokering can be done.

Let us examine some specific Web mining outcomes that will benefit e-commerce. With Web mining, one can form clusters of consumers. That is, members of group A are likely to buy toys (because they have children less than ten years old), and members of group B are likely to buy cosmetics (because they are teenage girls). Another outcome is an association, for example, if one sister prefers a certain brand of clothing, then other sisters will also prefer the same brand. On the consumer side, data mining may provide the following type of information: check out company X's Web site before purchasing the product from company Y.

As mentioned in Thuraisingham,[128] not only does data mining benefit e-commerce sites, as discussed in the previous paragraph, data mining also helps users find information on the Web. One e-commerce site manager mentioned to these authors that the major problem he encounters is users finding his e-commerce site. He has advertised in various magazines, but those who have no access to those magazines find it difficult to locate his site. One possible solution would be to have a third party agent make the connection between the site and the user. There are many agents on the Web who make connections between buyers and the sellers. For example, if someone wants to buy real estate, he contacts the agent. The agent associates him with the potential seller or a real estate agent.

Another solution would be to make the search engines more intelligent. Data mining can also help here. The data miner could take the requirements of the user and try to match the requirements to what is being offered by the e-commerce sites

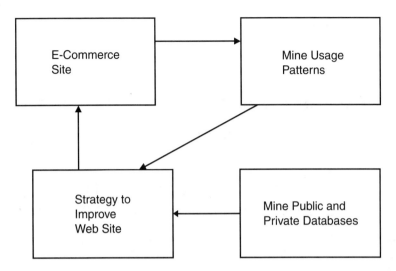

FIGURE 10.5 Web mining for e-commerce.

and connect the user to the right site. Work is just beginning in this area, and we still have a long way to go. Figure 10.5 illustrates data mining helping search engines. There are two approaches; one is to integrate the data miner with the search engine to create an intelligent search engine. This would be tight integration. The other approach is to have a data miner tool on the Web that interacts with multiple search engines so that there is loose integration between the search engine and the data miner.

Figure 10.6 illustrates a complete view of data mining for e-commerce. That figure illustrates the data miner helping the server carry out targeted marketing as well as helping the client locate the server. More information about data mining for e-commerce is given by Grasso et al.[47]

Another area of multimedia data mining important to e-commerce is distributed and collaborative data mining. With e-business, corporations will work together to carry out business operations. Even within a corporation, different divisions may have to work together. The various e-commerce distributed databases have to be mined to extract information. There are two approaches. One is to mine the individual data sources and then put the pieces together to form the big picture, as shown in Figure 10.7. The other is to carry out distributed data mining, as shown in Figure 10.8. Figure 10.9 illustrates two teams at different sites use collaboration and mining tools and mine the shared database.

Some aspects of distributed data mining were discussed by Thuraisingham,[127] and that topic is revisited in Section 10.5. The next chapter discusses multimedia for collaboration in more detail. Excellent research on these topics is being carried out by Guo et al.[50] at Imperial College in London.

Closely related to data mining is decision support. Decision support tools have to be integrated with e-commerce tools so that effective decisions can be made. Figure 10.10 illustrates e-commerce and decision support. Decision support tools can also be integrated with the data miners.

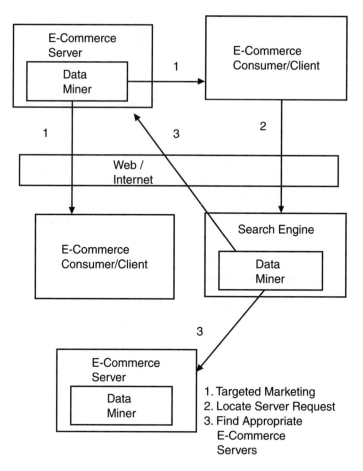

FIGURE 10.6 Multimedia data mining and e-commerce.

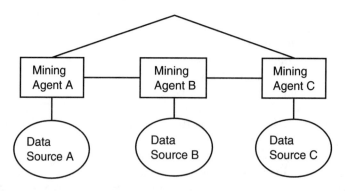

FIGURE 10.7 Collaborative data mining — approach I.

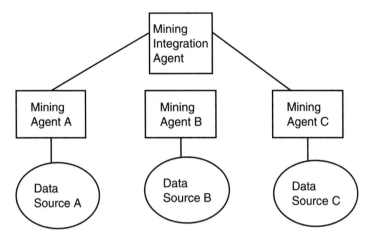

FIGURE 10.8 Collaborative data mining — approach II.

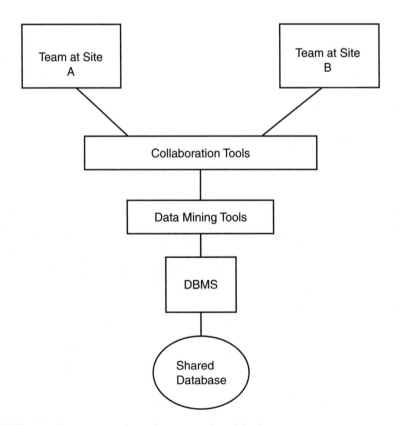

FIGURE 10.9 Teams conducting mining on a shared database.

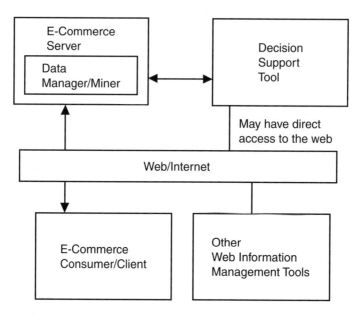

FIGURE 10.10 Decision support and e-commerce.

When more developments are made for multimedia data mining and the Web, we can expect better tools to emerge for Web mining, both to mine the data on the Web and to mine the usage patterns. We can expect to hear a lot about Web mining in the coming years.

10.5 AGENTS FOR MULTIMEDIA DATA MANAGEMENT AND MINING

The previous section briefly showed how agents can be used for collaborative data mining. Agents have a much broader use in multimedia data management and mining. As mentioned by Thuraisingham,[128] various views of agents have been proposed. That is, some say that agents are simply processes, and others believe that agents are Java applets. A third group feels that agents are processes that can jump from machine to machine and execute everywhere, and a fourth group says that agents are processes that have to communicate according to some well defined protocol. After examining the various definitions of agents, DiPippo et al.[40] give the following definition: [an agent is] "a computer system, situated in some environment, that is capable of flexible autonomous action in order to meet its design objectives."

The ideal is to get the right multimedia information at the right time to the users. This can be achieved either through a push model, where information is pushed to the user, a pull model, where the user goes out and gets the data, or a combination of both push and pull. Thuraisingham[128] discussed various models for communications, including the push and pull models. Essentially, it all comes down to information dissemination. That is, producers produce all kinds of multimedia information such as audio, vide, text, and images. This information has to be disseminated

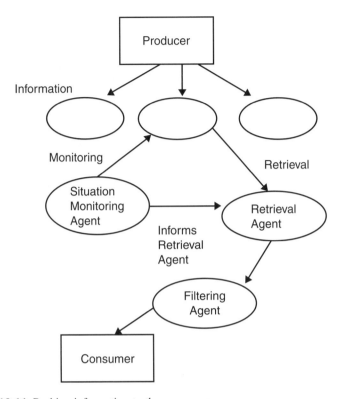

FIGURE 10.11 Pushing information to the consumer.

to the users in an appropriate manner. Now that we have examined various aspects of agents, we discuss the role of agents in information dissemination.

As stated by Thuraisingham,[128] agents could be locator agents that locate the resources, retrieval agents that retrieve data either by monitoring or when requested, situation monitoring agents that monitor for events, and filtering agents that filter unwanted information. All of these agents play a role in information dissemination. Figure 10.11 illustrates a scenario where situation agents monitor for information production, and this information is retrieved and filtered and then given to the consumer. Figure 10.12 illustrates a case where the consumer requests information, the locator agent locates the producers, and then the retrieval agent retrieves the information.

Information dissemination technologies have expanded due to the Web. The challenge is to get the information to the user without overloading the user. Since this is such a big challenge, we cannot expect this problem to be solved completely. However, developing technologies show a lot of promise, so information dissemination is enhanced.

10.6 DISTRIBUTED MULTIMEDIA DATA MINING

Thuraisingham[126] placed much emphasis on heterogeneous database integration and interoperability. Many enterprise applications require the integration of multiple data

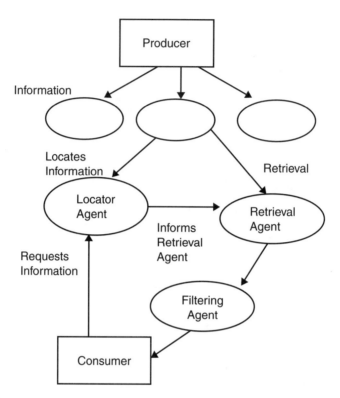

FIGURE 10.12 Pulling information from the producer.

sources and databases. We discussed the impact of multimedia data on integration in Part I. The distributed and heterogeneous multimedia data sources may need to be mined to uncover patterns. Furthermore, interesting patterns may be found across multiple databases. Mining heterogeneous and distributed data sources is a subject that has received little attention.

In the case of distributed multimedia databases, one approach is to have the data mining tool be part of the distributed processor (DP), where each DP has a mining component, as illustrated in Figure 10.13. That way, each data mining component can mine the data in the local database, and the DP can combine all the results. That would be quite challenging because the relationships between the various fragments of the relations or objects have to be maintained in order to mine effectively. Also, the data mining tool could be embedded into the query optimizer of the DQP, the distributed query processing component of the DP. Essentially, in this approach, the DP has one additional module that is a distributed data miner. We call this module a DDM, as shown in Figure 10.14.

Figure 10.15 illustrates an example of distributed data mining. Each DDM mines data from a specific database. These databases contain information on projects, employees, and travel. The DDMs can mine and get the following sample information: John and James travel together to London on project XXX at least ten times a year. Mary joins them at least four times a year.

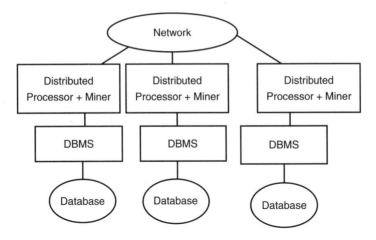

FIGURE 10.13 Distributed processing and mining.

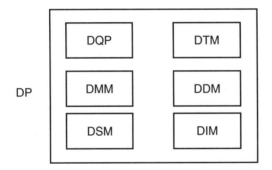

FIGURE 10.14 Modules of a DP for data mining.

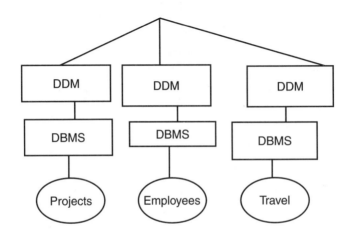

FIGURE 10.15 Example of distributed data mining.

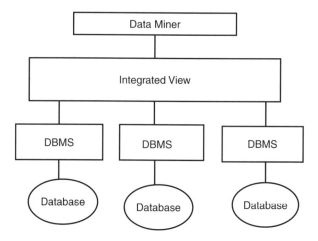

FIGURE 10.16 Data mining hosted on a distributed database.

An alternative approach is to implement the data mining tool on top of the distributed system. As far as the mining tool is concerned, the database is one monolithic entity. The data in this database has to be mined, and useful patterns have to be extracted, as illustrated in Figure 10.16.

In the case of heterogeneous data sources, we can either integrate the data and then apply data mining tools, as shown in Figure 10.17, or apply data mining tools to the various data sources and then integrate the results, as shown in Figure 10.18. Note that if the databases are integrated first, integration methods for interoperating heterogeneous databases are different than those for providing an integrated view in a distributed database. Some of these issues are discussed by Thuraisingham[126] (see also Reference 60). Furthermore, for each data mining query, one may first need to send that same query to the various data sources, get the results, and integrate

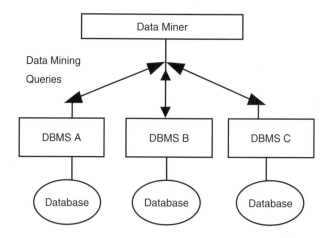

FIGURE 10.17 Data mining on heterogeneous data sources.

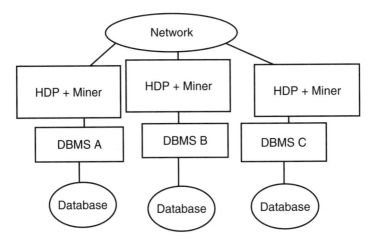

FIGURE 10.18 Mining and then integration.

the results, as shown in Figure 10.17. If the data is not integrated, then a data miner may need to be integrated with the heterogeneous distributed processor (HDP), as illustrated in Figure 10.18. If each data source is to have its own data miner, then each data miner acts independently. We do not send the same query to the different data sources because each data miner will determine how to operate on its own data. The challenge is to integrate the results of the various mining tools applied to the individual data sources so that patterns may be found across data sources.[31,37]

If one is to integrate the data sources and then apply the data mining tools, does he develop a data warehouse and mine the warehouse, or does he mine with inter-operating database systems? Note that in the case of a warehouse approach, not all of the data in the heterogeneous data sources are brought into the warehouse. Only decision support data are brought into the warehouse. If interoperability is used together with warehousing, then the data miner could augment both the HDP and the warehouse, as illustrated in Figure 10.19.

One can also use mediators to mine heterogeneous data sources. Figure 10.20 illustrates an example that assumes that there are general purpose data miners, and mediators are placed between the data miners and data sources. A mediator is also used to integrate the results from the different data miners.

Next, let us focus on legacy databases. One of the challenges is how to mine legacy databases. Can one rely on the data in these databases? Is it worth organizing and formatting this data, especially if it has to be migrated to newer systems? Is it worth developing tools to mine legacy databases? How easy is it to integrate legacy databases to form a data warehouse? There are some options. One is to migrate the legacy databases to new systems and mine the data in the new systems (see Figure 10.21). The second approach is to integrate legacy databases and form a data warehouse based on new architectures and technologies and then mine the data in the warehouse (see Figure 10.22). In general, it is not a good idea to directly mine legacy data, because this data could soon be migrated or incomplete and uncertain and, therefore, expensive to mine. Note that mining could also be used to reverse

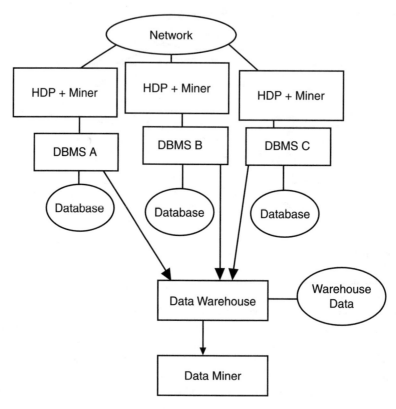

FIGURE 10.19 Mining, interoperability, and warehousing.

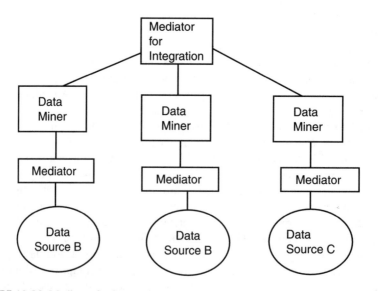

FIGURE 10.20 Mediator for integration.

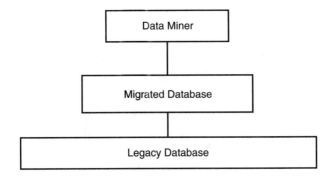

FIGURE 10.21 Migration and then mining.

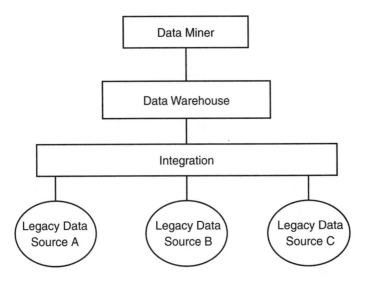

FIGURE 10.22 Mining legacy databases.

engineer and extract schemas from legacy databases (see Figure 10.23), and it may have been useful to handle the year 2000 problem. For a discussion of the year 2000 problem, refer to Thuraisingham.[126]

10.7 MINING AND METADATA

As discussed in Parts I and II, metadata is becoming a key technology for various tasks such as multimedia data management, data warehousing, Web searching, and now multimedia data mining. The focus of this section is on how metadata can be used for data mining. All of the discussion here also applies for multimedia data mining. Note that mining metadata may be especially important for multimedia data because, in many cases, mining metadata may be the only choice available for multimedia data.

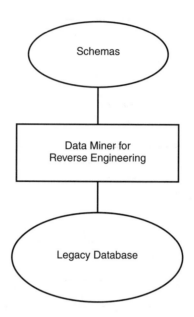

FIGURE 10.23 Extract schemas from legacy databases.

Metadata plays various roles in data mining. Metadata may guide the data mining process. That is, the data mining tool can consult the metadatabase and determine the types of queries to pose to the DBMS. Metadata may be updated during the mining process. For example, historical information as well as statistics may be collected during the mining process, and the metadata has to reflect the changes in the environment. The role of metadata in guiding the data mining process is illustrated in Figure 10.24.

There is also another aspect to the role of metadata, which is conducting data mining on the metadata. Sometimes the data in a database may be incomplete and inaccurate, and the metadata might contain more meaningful information. In such

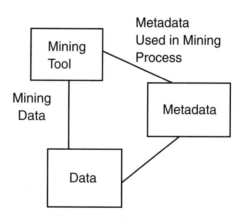

FIGURE 10.24 Metadata used in data mining.

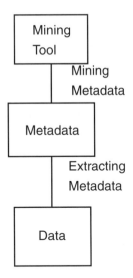

FIGURE 10.25 Metadata mining.

a situation, it may be more feasible to mine the metadata and uncover patterns. Mining metadata is illustrated in Figure 10.25.

There has been much discussion recently on the role of metadata for data mining.[84] There are many challenges involved. For example, when is it better to mine the metadata? What are the techniques for metadata mining? How does one structure the metadata to facilitate data mining? Researchers are working on addressing these questions.

Closely associated with the notion of metadata is that of a repository. A repository is a database that stores metadata, the mappings between various data sources when integrating heterogeneous data sources, information needed to handle semantic heterogeneity, for example, "ship X and submarine Y are the same entity," policies and procedures enforced, and information about data quality. So, the data mining tool may consult the repository to carry out the mining. On the other hand, the repository itself may be mined. Both of these scenarios are illustrated in Figure 10.26.

Metadata plays an important role in various types of mining. For example, in the case of mining multimedia data, metadata may be extracted from the multimedia databases and then used to mine the data. For example, as illustrated in Figure 10.27, the metadata can help extract the key entities from the text. These entities may be mined using commercial data mining tools. Note that in the case of textual data, metadata may include information such as the type of document, the number of paragraphs, and other information describing the document but not the contents of the document itself.

Metadata is also critical in the case of Web mining. Since there is so much multimedia information and data on the Web, mining this data directly could become quite challenging. Therefore, metadata may need to be extracted from the data, and then either mined or used to guide the mining process. This is illustrated in Figure 10.28.

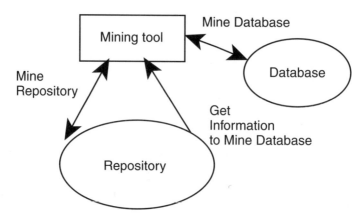

FIGURE 10.26 Repository and mining.

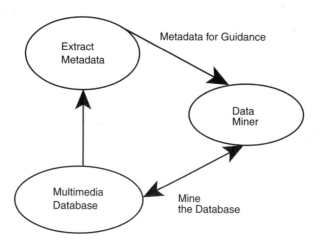

FIGURE 10.27 Metadata for multimedia mining.

Chapter 12 addresses security and privacy issues for multimedia data mining. We believe that policies and procedures will be a key issue in determining the extent to which the privacy of individuals is protected. These policies and procedures can be regarded as part of the metadata. Therefore, such metadata will have to guide the process of data mining so that privacy issues are not compromised through mining.

Metadata plays a crucial role in almost every aspect of mining. Even in the case of data warehousing, which we have regarded as a preliminary step to mining, it is important to collect metadata at various stages. For example, in the case of a data warehouse, data from multiple sources have to be integrated. As discussed in Thuraisingham,[127] metadata will guide the transformation process from layer to layer in building the warehouse (see Figure 10.29). Metadata will also help administrate the data warehouse. Also, metadata is used when extracting answers to the various queries posed.

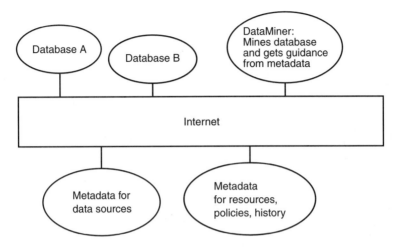

FIGURE 10.28 Metadata for web mining.

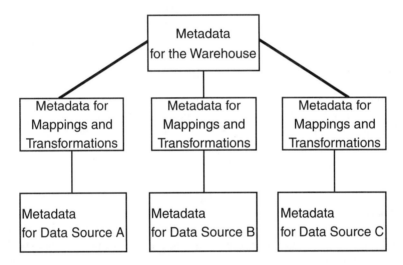

FIGURE 10.29 Metadata for transformations.

Since metadata is key to all kinds of databases, including relational databases, object databases, multimedia databases, distributed, heterogeneous, and legacy databases, and Web databases, one can envision building a metadata repository from the different kinds of databases and then mining the metadata to extract patterns. This approach is illustrated in Figure 10.30 and may be an alternative if the data in the databases are difficult to mine directly.

10.8 SUMMARY

This chapter has discussed the applications of multimedia data management and mining for the electronic enterprise. In particular, the applications of multimedia

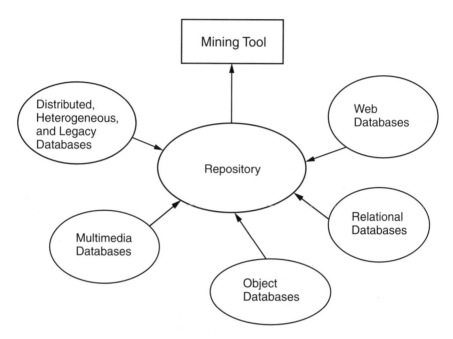

FIGURE 10.30 Metadata as the central repository for mining.

technology for the Web and e-commerce were described. We began with a general discussion of multimedia data processing issues. We then focused on both multimedia data management and multimedia data mining aspects for the Web and e-commerce. We also completed the key topics with discussions of agents for multimedia data mining, mining distributed multimedia databases, and metadata mining.

We believe that Web and e-commerce will be among the most powerful applications for multimedia information management, which includes both multimedia data management and mining. There is so much structured as well as unstructured data on the Web, that managing and mining that data will be critical for e-commerce. This chapter has discussed some of the possible future directions, but we still have a long way to go in this field.

11 Multimedia for Collaboration, Knowledge Management, and Training for the Web

11.1 OVERVIEW

Collaboration, knowledge management, and training are three important information management technologies for the Web. Multimedia information processing plays a major role in each of these technologies. Therefore, this chapter is devoted to a discussion of these technologies and the role of multimedia computing for these technologies for the Web and electronic enterprise.

The chapter begins by discussing collaborative computing and, in particular, multimedia data management for collaborative computing. Today, users are in different places collaborating on books, articles, and designs. These authors have collaborated with colleagues when writing papers for journals and conferences on the Internet. There are several tools that are emerging for collaboration, and these tools also work on the Web. While collaboration in the early 1990s mainly focused on collaborating within an organization, Web-based collaboration is now becoming a necessity.

Next, the chapter addresses knowledge management and multimedia for knowledge management. This is followed by a discussion of training and the use of multimedia data management for training and learning. Finally, we discuss the application of multimedia information processing for some other information technologies such as real-time processing, high performance computing, and visualization.

This chapter is organized as follows. Section 11.2 describes multimedia for collaboration. Multimedia for knowledge management is the subject of Section 11.3. Multimedia for training is the subject of Section 11.4. Section 11.5 deals with multimedia for some other information technologies. Finally, the chapter is summarized in Section 11.6.

11.2 MULTIMEDIA FOR COLLABORATION

11.2.1 OVERVIEW

Although the notion of computer supported cooperative work (CSCW) was first proposed in the early 1980s, only recently has any real interest been shown in this topic. Several research papers have now been published in collaborative computing, and prototypes/products have been developed. Collaborative computing enables individuals, groups of individuals, and organizations to work together in order to accomplish a task or a collection of tasks. These tasks may vary from participating in conferences to solving a specific problem to working on the design of a system. Specific contributions to collaborative computing include the development of team workstations (where groupware creates a shared workspace supporting dynamic collaboration in a work group), multimedia communication systems supporting distributed workgroups, and collaborative computing systems supporting cooperation in the design of an entity (such as an electrical or mechanical system).[2] Several technologies, including multimedia, artificial intelligence, networking and distributed processing, and database systems, as well as disciplines such as organizational behavior and human computer interaction, have contributed significantly towards the growth of collaborative computing.[2,64]

One aspect of collaborative computing of particular interest to the database community is workflow computing. Workflow is defined as the automation of a series of functions that comprise a business process such as data entry, data review, and monitoring performed by one or more people. An example of a process that is well suited for workflow automation is the purchasing process. Applications can range from simple user-defined processes such as document review to complex applications such as manufacturing processes. Original custom-made workflow systems developed over the past twenty years for applications such as factory automation were built using a centralized database. Many commercial workflow system products targeted for office environments are based on a messaging architecture. This architecture supports the distributed nature of current workteams. However, the messaging architecture is usually file based and lacks many of the features supported by database management systems such as data representation, consistency management, tracking, and monitoring. Although the emerging products show some promise, they do not provide the functionality of database management systems.

This chapter identifies the database systems technology issues to support collaborative computing in general and workflow computing applications in particular. Many of the ideas discussed here also apply for collaborative computing systems. However, we have limited the scope of our discussion by focusing mainly on workflow applications. There are two ways to design a data management system for a workflow application. One is a top-down approach, i.e., to design the entire application and then determine the type of data management system that is needed. The other is to focus only on the data management system. The data management system developed under the first approach would be specialized for the particular application, whereas the one developed under the second approach would be more general purpose. Both approaches are discussed here.

This section is organized as follows. First, some examples of database support for collaboration are given in Section 11.2.2. Architectural issues for workflow management systems are discussed in Section 11.2.3. General purpose database management system support for workflow applications is discussed in Section 11.2.4. In

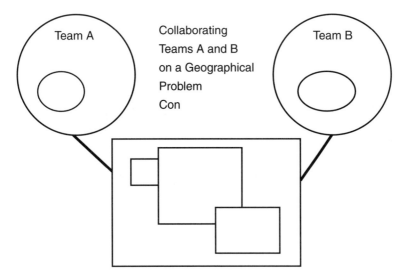

FIGURE 11.1 Collaboration example.

particular, data representation and manipulation as well as the role of metadata are discussed. The impact of the Web is the subject of Section 11.2.5.

11.2.2 SOME EXAMPLES

A collaborative computing system should enable multiple groups and teams from different sites to collaborate on a project. Figure 11.1 illustrates an example in which teams A and B are working on a geographical problem such as analyzing and predicting the weather in North America. The two teams must have a global picture of the map as well as any notes that go with it. Any changes made by one team are instantly visible to the other team, and both teams communicate as if they are in the same room.

To enable such transparent communication, data management support is needed. One can utilize a database management system to manage the data or some type of data manager that provides some of the essential features such as data integrity, concurrent access, and retrieval capabilities. In the above example, the database may consist of information describing the problem the teams are working on, the data that is involved, historical data, as well as metadata information. The data manager must provide appropriate concurrency control features so that when both teams simultaneously access the common picture and make changes, these changes are coordinated.

One possible scenario for the data manager is illustrated in Figure 11.2, where each team has its own local data manager, and a global data manager is there to maintain any global information, including the data and metadata. The local data managers communicate with the global data manager. The global data manager illustrated in this figure is at the logical level. At the physical level, the global data manager may also be distributed. The data managers coordinate their activities to provide features such as concurrency control, integrity, and retrieval.

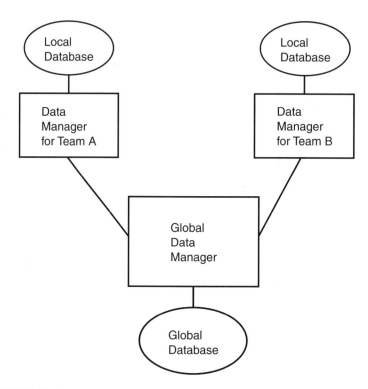

FIGURE 11.2 Database support.

11.2.3 ARCHITECTURAL SUPPORT FOR WORKFLOW COMPUTING

As stated above, workflow computing is a special case of collaborative computing. This section and the next two discuss various aspects of database support for workflow computing applications.

A database management system for a workflow application manages the database that contains the data required for the application. For a workflow application, the data could be purchase orders, requisitions, and project reports, among others. Like some of the other systems discussed in this book, there are various ways to integrate workflow systems with database management systems. We discuss two of the approaches here. In one approach, there is loose integration between the workflow management system and the database management system. This is illustrated in Figure 11.3. With this approach, one could use a commercial database management system for the workflow application. In the second approach, illustrated in Figure 11.4, there is tight integration between the workflow management system and the database management system. With this approach, the database management system is often a special purpose one.

The database management system could be centralized or distributed. All of the advantages and disadvantages for centralized and distributed database systems

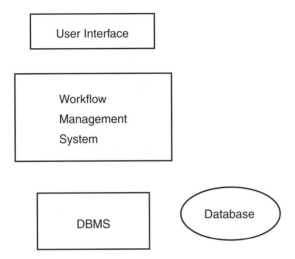

FIGURE 11.3 Loose integration between a workflow system and a DBMS.

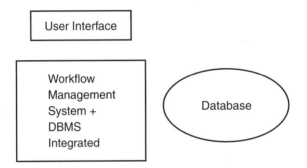

FIGURE 11.4 Tight integration between a workflow system and a DBMS.

discussed in Part I also apply for database management systems designed for workflow applications. In addition, the database management system should provide additional support for special transactions for workflow systems. Some database management system issues are discussed in Section 11.3. Note that one can also utilize distributed object management systems to encapsulate the components of a workflow management system.

As mentioned earlier, workflow computing is an aspect of collaborative computing. As illustrated in Figure 11.1, collaborative computing encompasses many features such as team members collaborating on a project, designing a system, and conducting a meeting. A collaborative application can be built on top of a workflow management system. The relationship between a database management system, a workflow management system, and a collaborative computing system is illustrated in Figure 11.5.

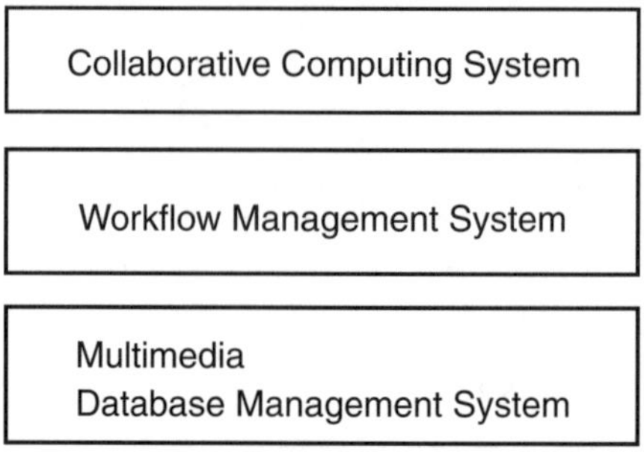

```
┌─────────────────────────────────────┐
│                                     │
│     Collaborative Computing System  │
│                                     │
└─────────────────────────────────────┘

┌─────────────────────────────────────┐
│                                     │
│     Workflow Management System      │
│                                     │
└─────────────────────────────────────┘

┌─────────────────────────────────────┐
│     Multimedia                      │
│     Database Management System      │
│                                     │
└─────────────────────────────────────┘
```

FIGURE 11.5 Relationship between systems.

11.2.4 MULTIMEDIA DATABASE SUPPORT
FOR WORKFLOW APPLICATIONS

11.2.4.1 Multimedia Database System Models and Functions

The database management functions for workflow applications will depend on the functionality requirements. These requirements can, in general, be divided into various categories, including data representation and data manipulation. The data representation requirements include support for complex data structures to represent (1) objects like documents, spreadsheets, and mail messages, (2) the workflow rules which determine the electronic routing of the objects, (3) tracking data which monitor the status of the objects, (4) the actions by users, (5) deadlines imposed on the actions, and (6) security constraints. Special formats for presenting the objects to users may also be needed. Data manipulation requirements include (1) querying and browsing the database, (2) managing the metadata, which describes the data in the database, (3) mechanisms for concurrent access to the data, such as modifying the workflow or the objects, (4) special view mechanisms for the users, (5) support for remote database access, and (6) integrating with the databases of other workgroup applications such as project management and document management.

The data model for a workflow application should support the data representation requirements. Various types of data models, including semantic models such as an object-oriented model and simple models consisting of nodes and links, are being investigated for this purpose. Also, a data model which accommodates different representation schemes may be needed for the different types of data. For example, the workflow rules may be represented in rule bases, and the documents, spread-sheets, and messages may be represented as complex objects. The data model should also provide the support to enforce integrity and security constraints, maintain different versions and configurations, and also represent the changes that an object goes through within a transaction. Also, if the objects are to be represented to the user in a different scheme, then mappings between the different schemes have to be stored. The data model should also be flexible to support ad hoc changes and schema evolution. The use of temporal constructs may be needed to represent historical and/or time-dependent data objects.

Since workflow applications are distributed in nature, both centralized as well as distributed architectures need to be examined for data management systems. For example, should all of the data such as workflow rules and objects be stored in a centralized location, or should the data be distributed across different workstations? If the architecture is distributed, then should it support a heterogeneous environment? From the discussion of the functionality requirements of workflow applications, described by Marshak,[81] it appears that the database should be fully integrated with the application. This means that the environment is most likely to be homogeneous. Note that if the workflow application has to be integrated with other applications such as project management, there may be a need to integrate the heterogeneous databases. Another issue is whether to store all of the data in the database. For example, changes that an object goes through during a transaction may be transient and need to be stored only through the duration of the transaction. In that case, the changes could be stored in temporary storage. If the changes have to be stored for historical purposes, then the database can be used.

The data management system should provide support for the data manipulation requirements. One of the main issues involved is developing a suitable model of concurrency control.[13,107] Since the goal is to enable the collaborators to share as much data as possible, a fine-grained granularity locking approach seems more appropriate than a coarse-grained one. Furthermore, alternatives to locking as a concurrency control mechanism as well as variations to the locking technique are being explored by researchers. In general, the model should be flexible so that a user can lock one part of an object while his peers work with other parts of the object. Furthermore, it is also desirable for the collaborators to be notified almost instantly if any part of an object is being modeled by a user, and if so, by whom. If this feature is not available, it will be difficult to maintain the consistency of an object. Concurrency control support is also needed for updating the workflow rules. In addition to developing suitable concurrency control algorithms, appropriate recovery mechanisms need to be developed to handle system failures and transaction aborts. Workflow transaction management is an active research area. This research is also being transferred into commercial products. Other functions of the system include querying and browsing, managing the metadata, enforcing appropriate access control policies, managing different versions and configurations, and monitoring the changes to an object within a transaction. In addition, the system must manage the links between the different users' shared objects. Since metadata plays an important role in collaborative computing, a discussion of this role is given separately.

In summary, workflow applications require efficient support for managing the database that may, possibly, be distributed. This section has identified some of the issues that need to be investigated. In particular, approaches to developing a data management system as well as issues concerning developing a data model, architecture, and modules for such a system which satisfy the functionality requirements were discussed. While the developments in database systems technology have contributed significantly to support new generation applications, applying these developments as well as generating new developments for collaborative computing applications in general, and workflow computing applications in particular, are the next challenges. Database system vendors are integrating their products with workflow systems. We expect significant developments to be made in workflow-based database management systems.

11.2.4.2 The Role of Metadata

Metadata plays a major role for collaborative computing and workflow applications. Metadata not only describes the data in a database, such as the schemas, it also contains other information such as access control rules, links between different objects, policies, information about the various teams collaborating, information about the various versions, and other historical information. For workflow computing applications, metadata includes information about projects, schedules, and other activities. Metadata should be used by the various teams to support their collaboration and also help provide a global synchronized picture of the problem being handled.

An appropriate model for the metadata is needed. Again, this model could be closely linked with the models used by the data managers or it could be completely independent. Various schema transformations between different representations are also included in the metadata. The metadata manager should provide support for querying and updating the metadata as well as giving advice to the various collaborating teams and guide the decision making process.

The metadata manager could be centralized or distributed. This design may depend on the architecture selected for the environment. If the database is distributed, for example, each team using its own local database, then there could be a metadata manager for each database. The various metadata managers have to communicate with each other.

Metadata research for workflow computing and collaborative applications is just beginning. However, with the recent emphasis on metadata for various applications including digital libraries, we can expect more results on metadata issues for collaborative computing.

11.2.5 IMPACT OF THE WEB ON COLLABORATION

The Web has increased the need for collaboration even further. Users now share documents and work on papers and designs on the Web. Corporate information infrastructures promote collaboration and sharing of information and documents. Therefore, the collaborative tools have to work effectively on the Web. While the Web promotes collaboration, collaboration also benefits the Web. That is, collaboration is a key information technology to enhance the Web. Therefore, the two technologies can benefit from each other.

The challenge is to use these tools to work effectively on the Web. The simple answer is to build Web interfaces. However, none of the tools were developed with the Web in mind. For example, that was the case with the database systems, and merely building Web access is the easy part. There are still many challenges, such as data formats, transactions, and metadata management on the Web. Such challenges are present for collaboration also. Some of the collaborative tools work with the understanding that the people involved are located in the same building. While for

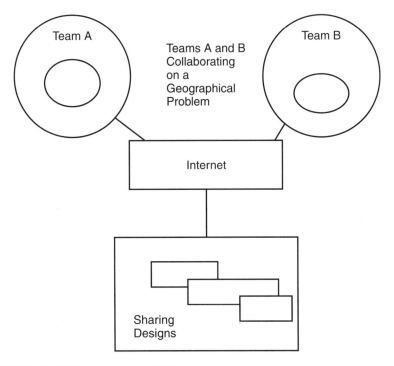

Team A

Teams A and B
Collaborating
on a
Geographical
Problem

Team B

Internet

Sharing
Designs

FIGURE 11.6 Collaboration on the Web.

corporate intranets this may not be a problem, this could pose some major problems for the Internet. Scalability of the tools is also an important issue. Typically, tools have been developed for tens of users. For the Web, these tools have to work for tens of thousands of users. Such requirements have to be taken into consideration. Figure 11.6 uses the same example discussed earlier in this section (geographical problem) and shows collaboration on the Internet.

We believe that collaboration and the Web will go hand in hand. In the future, collaboration tools will have to work with multimedia data. Therefore, multimedia data is addressed in the next section. Recent years have seen some interesting articles published on collaboration and the impact of the Web on collaboration tools.[64]

11.3 MULTIMEDIA FOR KNOWLEDGE MANAGEMENT

11.3.1 KNOWLEDGE MANAGEMENT CONCEPTS AND TECHNOLOGIES

Knowledge management is the process of using knowledge as a resource to manage an organization. It could mean sharing expertise, developing a learning organization, teaching the staff, learning from experiences, and collaboration. Essentially, knowledge management includes data management and information management. However, this view is not shared by everyone. Various definitions of knowledge management have been proposed. A good reference for knowledge management is

FIGURE 11.7 Knowledge management components.

Davenport.[32] Knowledge management is a discipline invented mainly by business schools. The concepts have been around for a long time. But the term knowledge management was coined as a result of information technology and the Web. In the collection of papers on knowledge management by Morey et al.,[88] knowledge management is divided into three areas, as shown in Figure 11.7. These are strategies such as building a knowledge company and hiring a staff of knowledge workers, processes such as techniques for knowledge management including developing a method to share documents and tools, and metrics that measure the effectiveness of knowledge management. The Harvard Business Review on knowledge management contains an excellent collection of articles describing creating a knowledge company, building a learning organization, and teaching people how to learn.[53] Organizational behavior and team dynamics play major roles in knowledge management.

Knowledge management essentially changes the way an organization functions. Instead of competition, it promotes collaboration. This means that managers have to motivate their employees to share ideas and collaborate by giving awards and other incentives. Team spirit is essential for knowledge management. People are often forced to impart knowledge because their jobs are on the line. They are reluctant to share expertise. This type of behavior varies from culture to culture. It is critical that managers eliminate this reluctance not by forcing the issue, but by motivating the staff and educating them as to all the benefits that can occur with good knowledge management practices.

Teaching and learning are two important aspects of knowledge management. Both the teacher and the student have to be given incentives. Teachers can benefit by getting thank-you notes and write-ups in the company newsletter. Students may be rewarded with certificates, monetary awards, and other similar gestures. Knowledge management also includes areas such as protecting the company's intellectual properties, job sharing, changing jobs within the company, and encouraging change in an organization. Effective knowledge management eliminates a dictatorial management style and promotes a more collaborative management style. Knowledge management follows a cycle of creating, sharing, and integrating knowledge, evaluating the performance with metrics, and then giving feedback to create more knowledge. This is illustrated in Figure 11.8. Variations of this cycle have been proposed in the literature.[86]

The major question involves defining knowledge management technologies. This is where information technology comes in. Artificial intelligence researchers have

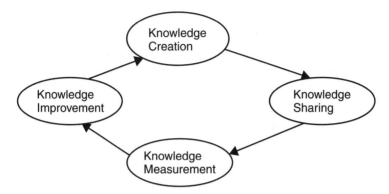

FIGURE 11.8 Knowledge management cycle.

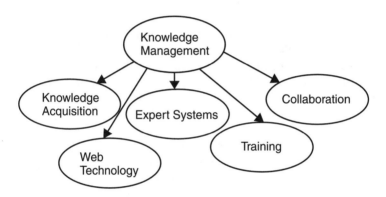

FIGURE 11.9 Knowledge management technologies.

carried out a considerable amount of research on knowledge acquisition. They have also developed expert systems. These are also knowledge management technologies. Other knowledge management technologies include collaboration tools, tools for organizing information on the Web, and tools for measuring the effectiveness of the knowledge gained, e.g., collecting various metrics. Knowledge management technologies essentially include data management and information management technologies. Figure 11.9 illustrates some of the knowledge management technologies. As can be seen, Web technologies play a major role in knowledge management. The impact of the Web is the subject of the next subsection.

11.3.2 The Role of Multimedia Computing

The main focus of this book is multimedia data management and mining. What is the impact of multimedia computing on knowledge management? Knowledge management is the process of effectively managing and sharing the resources of an organization. The resources may be presentations, papers, expertise, and speeches. This means that much of the data will be in the form of text, images, video, and

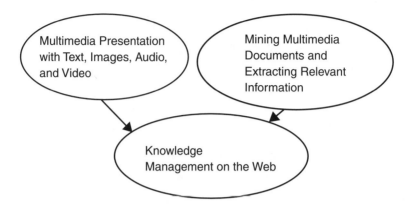

FIGURE 11.10 Multimedia computing and knowledge management.

audio. Therefore, one needs some kind of multimedia information management capability to manage the resources of the organization. Essentially, multimedia information management is the key to effective knowledge management.

Multimedia data mining is another important application for knowledge management. There is so much multimedia data in an organization that extracting useful information is critical. That is, the multimedia data has to be mined so that only relevant information is given to the user. That way, the members of an organization will get only the information that is relevant to them. Data mining will also explain various difficult concepts so that users can understand complex data. The role of multimedia computing for knowledge management is illustrated in Figure 11.10.

11.3.3 Knowledge Management and the Web

Knowledge management and the World Wide Web are closely related. While knowledge management practices have existed for many years, it is the Web that has promoted knowledge management. Remember, knowledge management is essentially building a knowledge organization. No technology is better than the Web for sharing information. One can travel around the world in seconds on the Web. As a result, a tremendous amount of knowledge can be gained by browsing the Web.

Many corporations now use intranets, which are the single most powerful knowledge management tool. Thousands of employees in a single organization are connected through the Web. Large corporations have sites all over the world, and employees are well connected with one another. E-mail can be regarded as one of the early knowledge management tools. Now there are many additional tools like search engines and e-commerce tools.

With the proliferation of Web data management and e-commerce tools, knowledge management will become an essential part of the Web and e-commerce. Figure 11.11 illustrates the knowledge management activities on the Web, such as creating Web pages, building e-commerce sites, sending e-mail, and collecting metrics on Web usage.

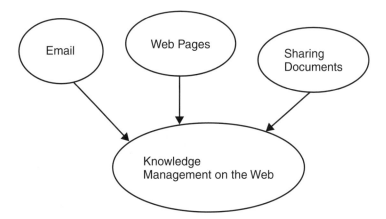

FIGURE 11.11 Knowledge management on the Web.

11.4 MULTIMEDIA FOR TRAINING

11.4.1 TRAINING AND DISTANCE LEARNING

Computer-based training, more popularly known as CBT, is a hot topic these days. CBT involves preparing course materials and posting them electronically so that trainees can learn at their own pace. Since the instructor is often not present, there are several user interface issues, and human computer interaction aspects come into play. One challenge is providing a personalized service to the trainees. For example, in a course on data management, financial workers may want additions on e-commerce, while defense workers may want information on data management for government applications. Instructors have to interview users, gather requirements, and prepare the material according to the requirements.

One aspect of CBT is training on the Web. This is also a form of distance learning. We now see several universities offering degrees on the Web based on distance learning. One challenge is not only preparing the material to satisfy the users but also delivering the material in a timely manner. Multimedia on the Web is an important technology for this application because live teaching may be desired at times. That is, while CBT is extremely useful, from time to time, students may want contact with an instructor who may be thousands of miles away. Distance learning is not restricted to within a country. It is now being implemented across continents. Figure 11.12 illustrates CBT on the Web. Several technologies have to work together, not just CBT but also multimedia, real-time processing, and all of the technologies for Web data management. We can expect to hear a lot about CBT on the Web and distance learning over the next several years.

11.4.2 MULTIMEDIA FOR TRAINING

The previous section discussed CBT. Multimedia information processing is essentially the key to CBT. Since CBT has to have all of the details clearly explained, text alone

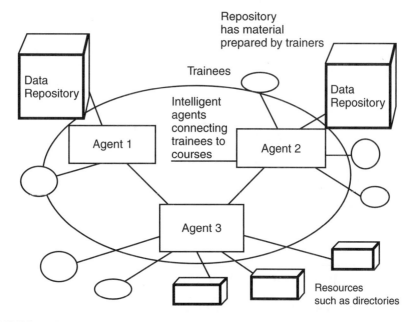

FIGURE 11.12 Computer-based training on the Web.

is not sufficient. Audio data management is necessary for students to listen to the instructor, and video data management is necessary so that students can see the instructor. The key is to create an experience that mimics the instructor actually being in the room. This means efficient multimedia information management techniques are critical.

In addition to supporting effective teaching, multimedia technologies can also help with presentation materials. That is, presentation materials may not only contain text, but also images, audio, and video data. We hear more and more about on-line training and Web universities. Multimedia information management will be critical for the successful operation of such novel teaching organizations. Figure 11.13 illustrates multimedia technology for training and e-learning.

11.5 MULTIMEDIA FOR OTHER WEB TECHNOLOGIES

11.5.1 OVERVIEW

The previous sections discussed the use of multimedia computing for collaboration, knowledge management, and training. These are three important applications for the Web. This section describes the application of multimedia computing for some other information technologies. Section 11.5.2 discusses real-time computing. Section 11.5.3 discusses visualization. Section 11.5.4 discusses quality of service aspects. Finally, some future directions are discussed in Section 11.5.5.*

---

* Note that topics such as visualization were also discussed in Chapter 8. We discuss it here for completion.

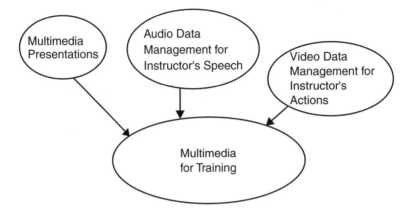

FIGURE 11.13 Multimedia for training.

11.5.2 REAL-TIME AND HIGH PERFORMANCE COMPUTING

Real-time computing is all about meeting timing constraints and deadlines. That is, transactions and queries have to meet timing constraints. Although fast computing is an important requirement for meeting constraints, it is not sufficient. For many applications, for example, in e-commerce, traders may want answers about stocks within five seconds, otherwise it will be too late. Essentially, these are hard real-time constraints. There are also soft real-time constraints, where it is desirable to meet timing constraints for as many transactions as possible.

The previous section discussed the need for real-time scheduling for multimedia applications. For example, it may be important to display voice and video together synchronously. Real-time operating systems, networks, and databases have to work together to meet timing constraints. More recently, a lot of work has been done on real-time middleware and especially object-oriented real-time middleware.[142] These real-time technologies now have to be effectively integrated into the Web. One promising development are the specifications for real-time Java. The idea is to develop technology to allow the Java scheduling mechanisms to meet the real-time constraints.[72] This will be a key direction for e-commerce where the timing constraints of transactions have to be met and many of the transactions may be written in Java. Integration of real-time and e-commerce technologies is in its infancy. We can expect to see a lot of progress there. Figure 11.14 illustrates real-time processing on the Web.

There has been much work done on integrating real-time computing with multimedia computing. For example, real-time scheduling techniques have been applied to synchronize different data types, as illustrated in Figure 11.15.

In a parallel database system, the various operations and functions are executed in parallel. While research on parallel database systems began in the 1970s, only recently are these systems being used for commercial applications. This is partly due to the explosion of data warehousing and data mining technologies where performance of query algorithms is critical.

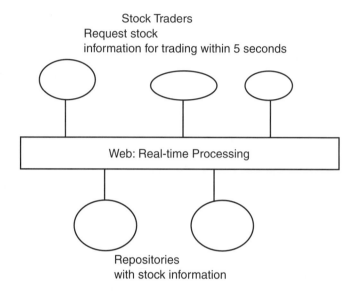

FIGURE 11.14 Real-time processing on the Web.

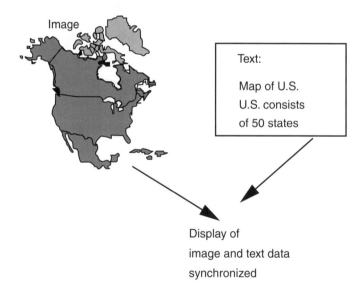

FIGURE 11.15 Real-time multimedia processing.

Let us consider a query operation which involves a join operation between two relations. If these relations are to first be sorted before the join, then the sorting can be done in parallel. We can take it a step further and execute a single join operation with multiple processors. Note that multiple tuples are involved in a join operation from both relations. Join operations between the tuples may be executed in parallel. Recently, work has been done on developing join algorithms for multimedia data. We need to examine the use of parallel processing techniques for these join operations.

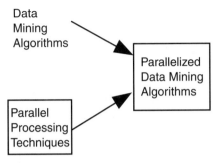

FIGURE 11.16 Parallel data mining.

Many commercial database system vendors are now marketing parallel database management technology. This is an area that will probably grow significantly over the next decade. One of the major challenges involved is the scalability of various algorithms for functions such as data warehousing and data mining. Also, we need to examine warehousing for multimedia databases.

Recently, parallel processing techniques have been examined for data mining. Many data mining techniques are computationally intensive. Appropriate hardware and software are needed to scale the data mining techniques. Database vendors are using parallel processing machines to carry out data mining. The data mining algorithms are put in parallel using various parallel processing techniques. This is illustrated in Figure 11.16.

Vendors of workstations are also interested in developing appropriate machines to facilitate data mining. This is an area of active research and development, and corporations such as Silicon Graphics and Thinking Machines (now part of the Oracle Corporation) have developed products. We can expect to see a lot of progress in this area during the next few years. With the advent of the Web, these various high performance computing tools have to work on the Internet. There are two types of tools needed. First, functions like mining have to be carried out on the Internet, and, therefore, parallel processing tools are needed to interface to the Web. Second, tools are also needed to make the Web more efficient and faster.

11.5.3 VISUALIZATION

Visualization technologies graphically display data in databases. Much research has been conducted on visualization, and the field has advanced a great deal, especially with the advent of multimedia computing. For example, data in databases could be rows and rows of numerical values. Visualization tools take the data and plot them in some form of a graph. The visualization models could be two-dimensional, three-dimensional, or more. Recently, several visualization tools have been developed to integrate with databases, and several workshops have been devoted to this topic.[135] An example illustration of integration of a visualization package with a database system is shown in Figure 11.17.

More recently, there has been a lot of discussion of using visualization for data mining. There has also been some discussion of using data mining to help the

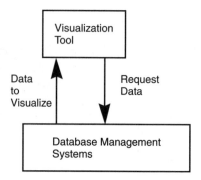

FIGURE 11.17 Database and visualization.

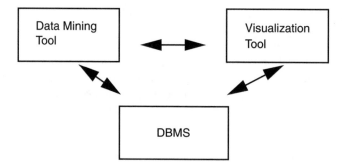

FIGURE 11.18 Interactive data mining.

visualization process. However, when considering visualization as a supporting technology, it is the former approach that is getting considerable attention.[48] As data mining techniques mature, it will be important to integrate them with visualization techniques. Figure 11.18 illustrates interactive data mining. In Figure 11.18, the database management system, visualization tool, and machine learning tool all interact with each other for data mining.

Let us re-examine some of the issues of integrating data mining with visualization. There are four possible approaches. One is to use visualization techniques to present the results that are obtained from mining the data in the databases. These results may be in the form of clusters or they could specify correlations between the data in the databases. The second approach applies data mining techniques to visualization. The assumption is that it is easier to apply data mining tools to data in the visual form. Therefore, rather than applying the data mining tools to large and complex databases, one captures some of the essential semantics visually and then applies the data mining tools. The third approach is to use visualization techniques to complement the data mining techniques. For example, one may use data mining techniques to obtain correlations between data or detect patterns. However, visualization techniques may still be needed to obtain a better understanding of the

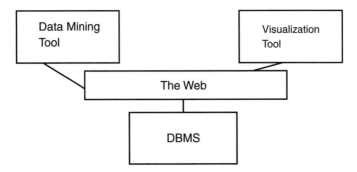

FIGURE 11.19 Visualization on the Web.

data in a database. The fourth approach uses visualization techniques to steer the mining process.

The various data visualization tools now have to work on the Web. For example, these tools need to access the various data sources on the Web and visualize them to help understand the data. This is illustrated in Figure 11.19.

11.5.4 QUALITY OF SERVICE ASPECTS

There are so many Web data management technologies, such as database management, security, fault tolerance, multimedia, mining, integrity, and real-time processing, that it will be a challenge to get all of them to work together effectively. For example, how do we guarantee that stock information meets the timing constraints for delivery to the trader and yet maintain one hundred percent security? That is very difficult. If we add the task of ensuring integrity of the data and techniques for recovering from faults and presenting multimedia data in a timely manner, the problem becomes nearly impossible to solve. So, what do we do? This is when quality of service, popularly known as QoS, comes in. It is almost impossible to satisfy all requirements all of the time. Therefore, QoS specifies policies for trade-offs. For example, if security and real-time are both constraints that have to be met, then perhaps in some instances, it is not absolutely necessary to meet all the timing constraints and we should focus on security instead. In other instances, meeting timing constraints may be crucial. As another example, consider multimedia presentation. In some cases, we can live with low resolution, while at other times we may need perfect pictures.

QoS has special relevance to multimedia computing. For example, in the case of video presentation, do we need smooth service all the time or can we tolerate breaks in service? In the case of image data, do we need full resolution? There again, QoS defines the primitives from multimedia presentations, and techniques to implement the QoS primitives specified in the policies need to be developed. In other words, not all applications will require the display of high quality data. In some cases, breaks in display of video data may be tolerated, while in other cases it may not. Figure 11.20 illustrates an example of low quality data in a normal mode and higher quality data in a crisis mode.

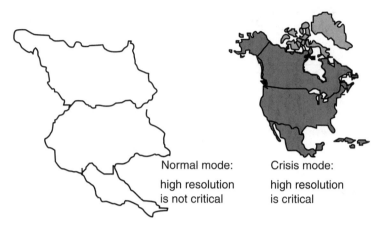

FIGURE 11.20 Quality of service.

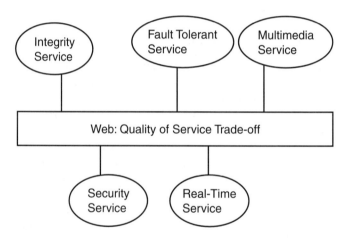

FIGURE 11.21 Quality of service trade-offs.

The challenges include specifying and implementing quality of service primitives. Formal approaches may help in this area. One needs to develop a model and an associated language for quality of service. The data model for multimedia database systems may also have to be extended with additional constructs.

Recently, there has been a lot of work done on QoS. But we are yet to find a model that takes into consideration all factors for quality of service. This is a difficult problem, yet with so many research efforts under way, we can expect to see progress. Essentially, the user specifies what he wants, and those requirements get mapped down to the database system, operating system, and the networking requirements. Figure 11.21 illustrates an approach to QoS on the Web. The ideas are rather preliminary and a lot remains to be done.

11.5.5 Some Future Directions

Thus far, this chapter has discussed the role of multimedia for various information technologies and services for the Web. Much of our discussion was focused on collaboration and multimedia technologies and services. We also briefly addressed training and distance learning, real-time processing, high performance technologies, and visualization aspects.

Many of the technologies we have discussed so far are about data and information management for the Web, and multimedia computing plays a major role. Another critical technology is knowledge management. One could argue that the whole area of knowledge management came about as a result of the Web. Knowledge management is all about capturing, storing, accessing, sharing, and even reusing the knowledge of organizations. A considerable amount of work on knowledge management has focused on sharing knowledge on the Web. As mentioned earlier, many corporations are using their internal information infrastructures as precious resources for knowledge management. In summary, knowledge management is a key technology for Web data management. Therefore, we give it special consideration in the next chapter.

Many technologies, such as data, information, and knowledge management as well as multimedia computing, are being used ultimately for managers, policymakers, and other people with authority to make effective decisions. Therefore, decision support is an important technology area for the Web because in the future, we can expect those managers and policymakers to access the Web and, based on the information they get, make effective decisions. Since decision support is also an important technology, we give it special consideration in the next chapter.

Finally, technology for accessing resources on the Web as well as processing those resources is critical for effective multimedia data management on the Web. The technology that is vital for these services is agent technology. There are different types of agents. Some agents locate resources, some carry out mediation, and some are mobile and execute in different environments.

Therefore, in addition to data, information, and knowledge management technologies, other technologies for multimedia information management and electronic commerce include security, collaboration, visualization, real-time processing, and training. There are several other multimedia technologies not addressed in this book. These include fault tolerance, mass storage, fuzzy systems and soft computing, and data administration. All of these technologies and services have to work together to make multimedia information management for the Web a success. Figure 11.22 illustrates how a distributed object management system can integrate the various technologies and services to provide effective multimedia data and information management for the Web.

11.6 SUMMARY

This chapter has discussed the applications of multimedia data management for key information technologies for the Web, including collaboration, knowledge management, and training. We also discussed other technologies such as real-time computing, visualization, and quality of service aspects.

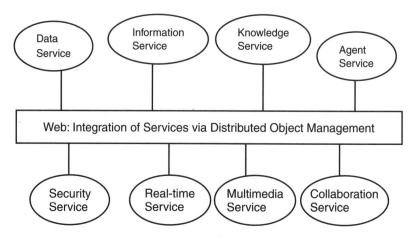

FIGURE 11.22 Integration of services on the Web.

These are not the only Web technologies central to multimedia data management and mining. Other technologies include interoperability, distributed processing, and agents, among others. A discussion of all of these technologies is beyond the scope of this book. Nevertheless, we have provided some directions for applying multimedia data management and mining for Web applications.

12 Security and Privacy Considerations for Managing and Mining Multimedia Databases

12.1 OVERVIEW

The number of computerized databases increased rapidly over the past three decades. The advent of the Internet as well as networking capabilities has made access data and information much easier. For example, users can now access large quantities of information in a short period of time. As more and more tools and technologies are being developed to access and use the data, there is also now an urgent need to protect the data. Many government and industrial organizations have sensitive and classified data that has to be protected. Various other organizations such as academic institutions also have sensitive data about their students and employees. As a result, techniques for protecting the data stored in databases are urgently needed.

Over the past three decades, various developments have been made for securing the databases.[128] In the 1990s, with the advent of new technologies such as digital libraries, the World Wide Web, and collaborative and multimedia computing systems, there was much interest in security, not only in government organizations, but also in commercial industry.

This chapter focuses on security for Web-based multimedia databases. We first provide an overview of the various issues of security and privacy for databases. For example, data on the Web has to be secured. However, with data mining tools, privacy may be compromised. The issues surrounding this topic are the subject of section 12.2. The impact of multimedia data on security and integrity are the subject of section 12.3. The chapter is summarized in section 12.4. For a detailed discussion of secure data management, refer to Ferrari and Thuraisingham.[143]

12.2 SECURITY AND PRIVACY FOR THE WEB

12.2.1 OVERVIEW

While security and privacy considerations for multimedia databases are our main consideration, we need to discuss security and privacy issues in general before we can examine the impact of processing multimedia data. Therefore, this section discusses general aspects of security and privacy for the Web, all of which are relevant to handling multimedia data.

While data mining has many applications in improving data quality and security,* there is also a dangerous side to mining, since it may be a serious threat to the security and privacy of individuals. This is because data mining tools are available on the Web, and even naive users can apply these tools to extract information from the data stored on the Web and, consequently, violate the privacy of various individuals.

One of the challenges of securing databases is the inference problem. Inference is the process of users posing queries and deducing unauthorized information from the legitimate responses they receive. This problem has been heavily discussed over the past two decades. However, data mining makes this problem worse. Users now have sophisticated tools that they can use to get sensitive data and deduce patterns. Without data mining tools, users would have to be fairly sophisticated in their reasoning to be able to deduce information from such sensitive data. That is, data mining tools make the inference problem quite dangerous.

Data mining approaches such as Web mining also seriously compromise the privacy of the individuals. One can obtain all kinds of information about a person in a short space of time by browsing the Web. Security for digital libraries, Internet databases, and electronic commerce is the subject of much research. Data and Web mining make this problem even more dangerous. Therefore, protecting the privacy of individuals is a major consideration.

This section discusses both the inference problems of data mining as well as privacy issues. In addition, a general discussion of some security measures for the Web and e-commerce is also provided. More details are given by Thuraisingham.[128] Section 12.2.2 provides an overview of the inference problem to give the reader some background and discusses approaches to handling specifically the inference problem that arises from data mining. Since data can be warehoused on the Web, we also describe data warehousing, inference, and security in Section 12.2.3. Since inductive logic programming is of interest to us, Section 12.2.4 discusses inference control through the use of inductive logic programming.** Note that there are various efforts to integrate logic programming with programming languages such as Java. Therefore, we can expect such extended logic programming systems to be used to provide security on the Web. Section 12.2.5 discusses privacy issues. These privacy issues also depend on policies and procedures enforced. That is, technical, political, and social issues play a role here. Then, Section 12.2.6 provides an overview of security measures on the Web.

12.2.2 BACKGROUND ON THE INFERENCE PROBLEM

Inference is the process of posing queries and deducing unauthorized information from the legitimate responses received. For example, the names and salaries of individuals may be unclassified, while taken together, they are classified. Therefore,

---

* Note that data mining can be used to improve efficiency, quality of data, and marketing and sales, and has many more benefits. Furthermore, even in the case of security problems, data mining tools can be used to detect abnormal behavior and intrusions in the system. Data mining also has many applications in detecting fraudulent behavior.

** For a discussion of inductive logic programming, refer to the private communication with the DSV Laboratory personnel[38] and to the summer school course given in Prague in 1998.[70]

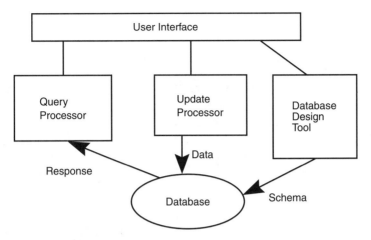

FIGURE 12.1 Addressing inference during query, update, and database design.

one could retrieve names and employee numbers and then later retrieve the salaries and employee numbers, and make the associations between names and salaries.

In the early 1970s, much of the work on the inference problem was on statistical databases. Organizations such as the census bureau were interested in this problem. However, in the mid 1970s and 1980s, the United States Department of Defense started an active research program on multilevel secure databases, and research on the inference problem (see, for example, the Air Force Summer Study Board Report[10]) was conducted as part of this effort. The pioneers included Morgenstern,[89] Thuraisingham,[113] and Hinke.[54]

We have conducted extensive research on this subject and worked on various aspects. In particular, it was shown that the general inference problem was unsolvable by Thuraisingham,[115] and then approaches were developed to handle various types of inference. These approaches included those based on security constraints as well as those based on conceptual structures.[119,121,123] They handled the inference problem during database design, query, and update operations (see the scenario in Figure 12.1). Furthermore, logic-based approaches were also developed to handle the inference problem.[116]

Much of the earlier research on the inference problem did not take data mining into consideration. With data mining, users now have tools to make deductions and patterns, which could be sensitive. Therefore, the next section addresses inference problems and data mining. It also includes some information on data warehousing and inference.

12.2.3 MINING, WAREHOUSING, AND INFERENCE

First, let us give an example where data mining tools are applied to cause security problems. Consider a user who has the ability to apply data mining tools. This user can pose various queries and infer sensitive hypotheses. That is, the inference problem occurs via data mining, which is illustrated in Figure 12.2. There are various ways to handle this problem. Given a database and a particular data mining tool,

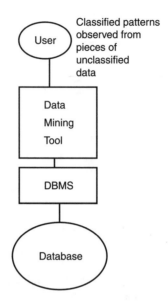

FIGURE 12.2 Inference problem.

one can apply the tool to see if sensitive information can be deduced from the unclassified information legitimately obtained. If so, then there is an inference problem. There are some issues with this approach, such as, it involves applying only one tool. In reality, the user may have several tools available to him. Furthermore, it is impossible to predict all the possible ways for the inference problem to occur. Some of the security implications are discussed by Clifton and Marks.[22]

Another solution to the inference problem is to build an inference controller that can detect the motives of the user and prevent the inference problem from occurring in the first place. Such an inference controller would lie between the data mining tool and the data source or database, possibly managed by a DBMS. This is illustrated in Figure 12.3. Discussions of security issues for data warehousing and mining can be found in Thursaisingham.[125]

Clifton[25] has also conducted some theoretical studies on handling the inference problems that arise from data mining. His approach is the following. If it is possible to cause doubts in the mind of the adversary that his data mining tool is not a good one, then he will not have confidence in the results. For example, if the classifier built is not a good one for data mining through classification, then the rules produced cannot have sufficient confidence. Therefore, the data mining results also will not have sufficient confidence. What are the challenges in making this happen? That is, how can we ensure that the adversary will not have enough confidence in the results? One approach is to give only samples of the data to the adversary so that he cannot build a good classifier from these samples (Figure 12.4 illustrates this scenario). The question, then, is what should the sample be? Clifton has used classification theory to determine the limits of what data can be released. This work is still preliminary. There have been some concerns about this approach, as one could give multiple samples to different groups, and the groups could then work together to build a good

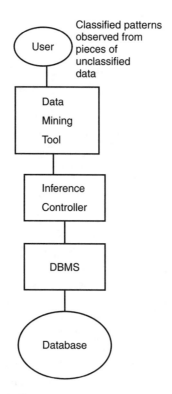

FIGURE 12.3 Inference controller.

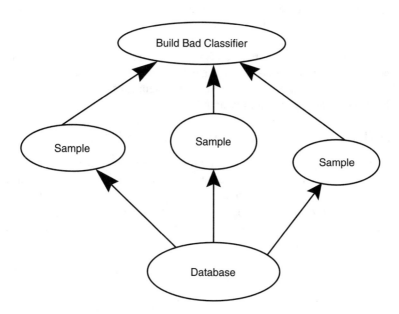

FIGURE 12.4 Approach to mining and inference.

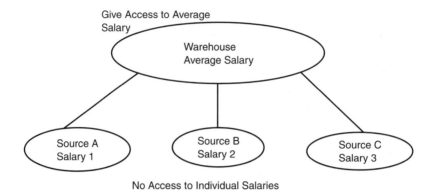

FIGURE 12.5 Warehousing and inference.

classifier. The answer to that, however, is that one needs to keep track of what information is given out. One way to handle the inference problem is not to give out any samples. But this could mean denial of service. That is, data could be withheld when it is definitely safe to do so.

Next, let us focus on data warehousing and inference. Some security issues for warehouses are addressed by Thuraisingham.[125] First of all, security policies of the different data sources that form the warehouse have to be integrated to form a policy for the warehouse. This is not a straightforward task, as one has to maintain security rules during the transformations. For example, one cannot give access to an entity in the warehouse, while the same person cannot have access to that entity in the data source. Next, the warehouse security policy has to be enforced. In addition, the warehouse has to be audited. Finally, the inference problem also becomes an issue here. For example, the warehouse may store average salaries. A user can access average salaries and then deduce the individual salaries in the possibly sensitive data sources (see the scenario in Figure 12.5), and, therefore, the inference problem becomes an issue for the warehouse. To date, little work has been reported on security for data warehouses as well as the inference problem for the warehouse. This is an area that needs much additional research.[67-69]

12.2.4 INDUCTIVE LOGIC PROGRAMMING (ILP) AND INFERENCE

The previous section discussed data mining and the inference problem. In our research, we have used deductive logic programming extensively to handle the inference problem. We have specified what we have called security constraints (see, for example, Thuraisingham[121]) and then augmented the database system with an inference engine, which makes deductions and determines if the constraints are violated. That is, the inference engine, by using the constraints, determines if the new information deduced causes security problems. If so, the data are not released.

The question is, can this approach be used to control inferences using inductive logic programming (ILP) techniques? Note that with inductive logic programming, one can infer rules from the data. That is, from various parent–children and grand-parent–grandchildren relationships, one can infer that the parent of a parent is a

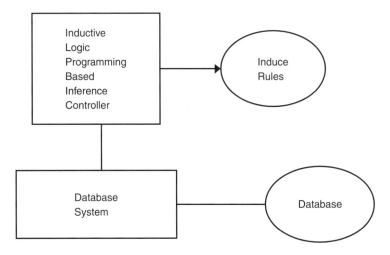

FIGURE 12.6 Inference controller based on ILP.

grandparent. Figure 12.6 is a possible architecture for such an inference controller based on inductive logic programming. This inference controller is based on inductive logic programming. It queries the database, gets the responses, and induces the rules. Some of these rules may be sensitive or lead to giving out sensitive information. Whether the rule is sensitive or can lead to security problems is specified in the form of constraints. If a rule is sensitive, then the inference controller will inform the security officer that the data has potential security problems and may have to be reclassified.

In other words, the inference controller suggested here does not operate on run time. As we have mentioned, it is very difficult to handle all types of data mining tools and prevent users from getting unauthorized responses to queries. What the inference controller does is give advice to the security officer regarding potential problems with the data and safety of the data. Some of the issues regarding inductive inference, which is essentially the technique used in ILP,* to handle the inference problem in secure databases are given by Thuraisingham.[117]

12.2.5 PRIVACY ISSUES

At the IFIP (International Federation for Information Processing) working conference on database security in 1997, the group began discussions on privacy issues and the role of the Web, data mining, and data warehousing.[68] This discussion continued at the IFIP meeting in 1998, and it was felt that the IFIP group should monitor the developments made by the security working group of the World Wide Web Consortium. The discussions that took place included those based on technical, social, and political aspects (see Figure 12.7). This section examines all of these aspects.

---

* For a discussion of logic programming, refer to Lloyd.[80]

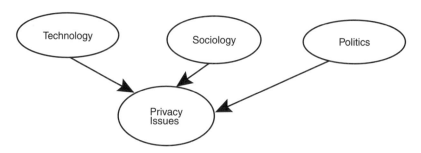

FIGURE 12.7 Privacy issues.

With the World Wide Web, there is now an abundance of information about individuals that one can obtain within seconds. This information can be obtained through mining or just by information retrieval. Therefore, one needs to enforce controls on databases and data mining tools. This is a very difficult problem especially with respect to data mining, as discussed in the previous section. In summary, one needs to develop techniques to prevent users from mining and extracting information from data, whether the data are on the Web or on servers. This goes against all we have said about mining in the previous chapters. That is, we have portrayed mining as something that is critical for users to have so they can get the right information at the right time as well as extract patterns previously unknown. That is all true. However, the information should not be used in an incorrect manner. For example, based on information about a person, an insurance company may deny insurance or a loan agency can deny loans. In many cases, these denials may not be legitimate. Therefore, information providers have to be very careful about what they release. Also, data mining researchers have to ensure that security aspects are addressed.

Next, let us examine the social aspects. In most cultures, privacy of individuals is important. However, there are certain cultures where it is impossible to ensure privacy. Reasons for this could be related to political or technological issues or the fact that people have been brought up believing that privacy is not critical. There are places where people divulge their salaries without thinking twice about it, but, in many countries, salary information is very private and sensitive. It is not easy to change cultures overnight, and, in most cases, preserving cultures is very important. So, what overall effect does this have on data mining and privacy issues? There is no answer to this yet as we are only beginning to examine cultural issues and their relevance to data mining.

Next, let us examine the political and legal aspects involved, including policies and procedures. What sort of security controls should one enforce for the Web? Should these security policies be mandated or discretionary? What are the consequences of violating the security policies? Who should be administering these policies and managing and implementing them? How is data mining on the Web impacted? Can one control how data are mined on the Web? Once we have made technological advances regarding security and data mining, can we enforce security controls on data mining tools? How is information transferred between countries? Again, we have no answers to these questions. We have, however, begun discussions. Note that some of the issues mentioned here are related to privacy and data mining, and others are related to privacy in general.

We have raised some interesting questions on privacy issues. As mentioned earlier, data mining is a threat to privacy. The challenge is to protect privacy but at the same time preserve all the great benefits of data mining. At the 1998 Knowledge Discovery in Databases conference, there was an interesting panel on privacy issues for Web mining. It appears that the data mining and security communities are interested in security and privacy issues. Much of the focus of that panel was on legal issues.[74]

12.2.6 SOME SECURITY MEASURES

So far, we have mainly dealt with privacy issues. Security is also a major consideration for the Web. The essentials of e-commerce security are Web security, for which there are three components in general: secure client, secure server, and secure network.

In an earlier work, Thuraisingham[128] discussed network security issues. The various network protocols have to be secured. In addition, the basic transmission has to be secure. Encryption provides this type of security. Data is encrypted at the sender's side and decrypted at the receiver's side. The main issue here is maintaining the encryption keys. Various techniques such as private key encryption, public key encryption, and certification methods have been used and were discussed by Ghosh.[46] In addition to network protocol security, the Web protocols, such as HTTP, have to be secure.

Traditional client–server security methods are used to ensure that clients and servers are secure on the Web. Furthermore, because of the nature of the Web and e-commerce transactions, additional security measures are needed. For example, user A may want to transfer funds from his account to user B's account. There could be a Trojan horse in the system transferring the funds to the account of user C instead. Now, if A and B are multinational corporations, millions of dollars could be lost. We hear about such security breaches all the time in the news. Various secure payment protocols and transactions methods have been developed to limit these types of breaches. Figure 12.8 illustrates the security layers for the Web.

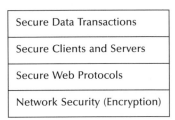

FIGURE 12.8 Secure protocols stack.

12.3 SECURE MULTIMEDIA DATA MANAGEMENT CONSIDERATIONS

12.3.1 ACCESS CONTROL AND FILTERING ISSUES

Note that the previous section discussed various aspects of security and privacy. In particular, it dealt with privacy violations that result from data mining as well as security for the Web. More details on security and privacy related to the Web are given by Thuraisingham.[128]

Since multimedia data management and mining is the major focus of this book, we are concerned with the impact of multimedia computing on security and privacy issues addressed in this chapter. In fact, whether the data is multimedia or not, we need to be concerned about all of the issues discussed in Section 12.2. However, multimedia data requires additional considerations. This section addresses some of those.

As in the case of data mining, multimedia data mining can also result in privacy violations. For example, one could mine text data and gather information about individuals that is not authorized. That is, text data provide more opportunities to violate privacy because text may contain more information. However, one needs to mine the text and get useful information, which is the challenge. The same is true for image, audio, and video data.

Another challenge with multimedia data is enforcing appropriate access control. The question is, what level of granularity should be provided? That is, should one have access control based on frames or the entire video clip? For example, user John does not have access to frames 1001 to 2000 but does have access to frames 2001 to 3000. This is frame-based access control, illustrated in Figure 12.9. Currently, the only other option is to give John access to the entire video clip. The challenge is to enforce finer levels of granularity for access control.

One of the difficulties today is protecting children from accessing inappropriate material on the Internet. This is especially true for image and video data. It is often harder to read the text, while it is much easier to look at images and video. Therefore, we need filtering techniques to ensure that only appropriate material is displayed to children. Enforcing such access controls is not only a technical challenge, but also a legal challenge. For example, we need to address issues such as the child's right to privacy.

Essentially, we feel that while many of the security challenges are the same for multimedia as well as non-multimedia data, multimedia data raises additional challenges because of the nature of the data and the Web. That is, the Web makes it

John does not have access to video frames 1001 to 2000	John has access to video frames 2001 to 3000	John does not have access to video frames 3001 to 4000

FIGURE 12.9 Access control to video frames.

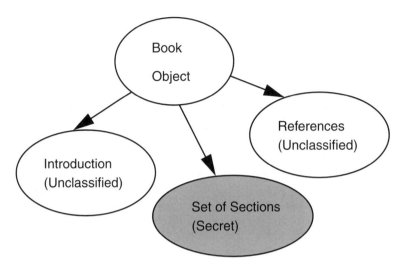

FIGURE 12.10 Example of a multilevel multimedia document.

easier to violate the privacy of individuals and makes the enforcement of security more difficult; one needs extensive research in this area. We believe that technologists have to work together with policy makers, psychologists and lawyers.

Multilevel security is also needed for certain multimedia applications. Some work has been reported on multilevel security.[118] For example, parts of a document might be classified, while other parts may be unclassified.[114] The representation of a multilevel multimedia document is illustrated in Figure 12.10. The issue here is for users to read only the documents classified at or below their clearance level.

12.3.2 DATA QUALITY AND INTEGRITY ISSUES

Data integrity and security are often discussed together. Integrity techniques ensure the accuracy of the data. Essentially, maintaining data integrity includes support for data quality, integrity constraint processing, concurrency control and recovery for multi-user updates, and accuracy of the data on output. Data mining may be used to ensure the integrity and quality of the data. That is, data mining techniques may determine if there is bad data. There are also many other issues that need to be considered, especially for multimedia data. Some of them are discussed in this section.

First of all, how can one guarantee accuracy of data? For example, in the case of video data, who filmed the data? Where did the data come from? How accurate is the data? What is the resolution of the data? In the case of text, is the data accurate? Is the data spelled correctly? Is the grammar correct? In the case of images, what are the quality of service primitives to display the data? How long has the data been archived? Has the data been tampered with? While some of these questions are common to both multimedia as well as non-multimedia data, there are many additional issues that need to be addressed for multimedia data. Chapter 4 discussed data quality and integrity as part of the discussion of metadata. There are still many issues that need to be addressed. We also need tools to check the quality of multimedia data.

12.4 SUMMARY

This chapter was devoted to the important area of security and privacy related to the Web as well as privacy issues for data mining. While there have been efforts to apply data mining to handling security problems such as intrusion detection (see, for example, Clifton[26]), this chapter focused on the negative effects of data mining. In particular, we discussed the inference problem that can result due to mining, as well as ways of compromising privacy especially due to Web data access. Finally, we discussed some general security measures for the Web.

The second part of this chapter dealt with multimedia considerations since the focus of this book is on managing and mining multimedia databases. All of the security and privacy concerns apply for multimedia as well as non-multimedia databases. However, with multimedia data, we have additional considerations such as controlling access to parts of images and video data. We also briefly addressed data quality and integrity issues for multimedia databases.

Security for multimedia databases is still a new area. Although Thuraisingham reported some work on multilevel security for multimedia databases,[118] there are still several aspects such as access control issues that need to be resolved. This chapter has only briefly addressed multimedia security issues. We believe that there is still a lot to be done in that field.

13 Standards, Prototypes, and Products for Multimedia Data Management and Mining

13.1 OVERVIEW

The previous chapters have provided some information on multimedia data management and multimedia data mining and how they may be applied to the electronic enterprise and e-business. We discussed aspects of multimedia data modeling, architectures, query processing storage management, data mining on text, images, audio, and video, and the impact of the Web. Many of the concepts addressed in this book have been influenced by the standards activities as well as the research prototypes. This, in turn, influences the commercial products. Researchers and standards organizations examine products, figure out ways to make improvements, and develop more prototypes and standards. Therefore, this is a cyclic operation.

This chapter examines the supporting activities for advancing multimedia data management and mining for the electronic enterprise. We begin with a discussion of standards. We discuss various aspects of standardization from query language to data mining. We then discuss the prototypes and products for multimedia data management. Finally, we discuss prototypes and products for multimedia data mining. The organization of this chapter is as follows. Section 13.2 describes standards activities. Multimedia data management prototypes and products are discussed in Section 13.3. Section 13.4 describes prototypes and products for multimedia data mining. The Chapter is summarized in Section 13.5. Figure 13.1 illustrates the three topics addressed in this chapter.

We are not endorsing any of the products discussed. We discuss a prototype or product mainly because of our familiarity with it. It does not mean that that prototype or product is superior. Furthermore, due to the rapid developments in the field, the information about these products may soon be outdated. Therefore, the reader is urged to take advantage of the various commercial and research material available on these products.

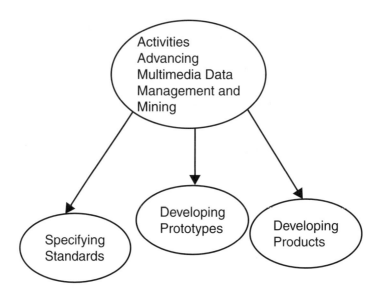

FIGURE 13.1 Advancing multimedia data management and mining.

13.2 STANDARDS

13.2.1 OVERVIEW

Standards play an important role in data management. Standards efforts have been underway in many areas including schema specification, data modeling, and query languages. More recently, with the advent of the Web, several additional standards have emerged for data management and interoperability. Examples include standards for object request brokers as well as standards for document representations such as XML.[93] As the data become more complex, additional standards are needed. There are various efforts underway to manage multimedia data. This section discusses a few of them.

We start with a discussion of query language standards followed by discussions of XML and ontologies. We then discuss standards for storage management. Finally, data mining standards as well as interoperability standards will be discussed. For many of the standards, we discuss the general concepts, point to references, and also discuss what is being done for multimedia data.

The organization of this section is as follows. Query language standards are the subject of Section 13.2.2. XML is addressed in Section 13.2.3. Ontologies are discussed in Section 13.2.4. Storage standards are the subject of Section 13.2.5. Data mining standards are described in Section 13.2.6. Interoperability standards will be the subject of Section 13.2.7. Finally, we discuss some other possible areas for standardization in Section 13.2.8. Figure 13.2 illustrates the various aspects of standards.

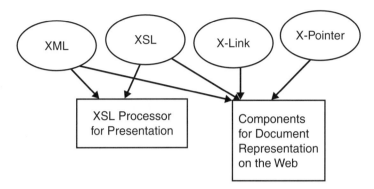

FIGURE 13.2 XML for the Web.

13.2.2 QUERY LANGUAGE

Various languages to manipulate the relations have been proposed. Notable among these languages is the ANSI Standard SQL (structured query language). This language is used to access and manipulate data in relational databases.[110] There is wide acceptance of this standard among database management system vendors and users. It supports schema definition, retrieval, data manipulation, schema manipulation, transaction management, integrity, and security. Other languages include the relational calculus first proposed by the Ingres project at the University of California at Berkeley.[32]

SQL is used both as a data manipulation as well as a data definition language. That is, SQL can be used to define the schemas and tables as well as manipulate the data for retrieval. The operations for data definition include create, delete, and insert. The operations for data manipulation include statements of the form, "Select from table where some condition is satisfied." This select statement is used to retrieve data using just one relation or multiple relations.

With multimedia data, one may use SQL for selecting elements from an object relation. For example, to retrieve the content portion of the multimedia object described in the previous section, an SQL statement would look like the following:

SELECT FROM TABLE

WHERE CONTENT = XXX

Here, table is the relation name and content is the string XXX. However, with image and video data, we need to extend the language to support image and video content. That is, we need a way to specify an image, such as the U.S. map, or a video, such as a particular film script. One way to do this is to express the content with a string, and then store the content information in the metadata. That is, the string may represent the name of a frame, while the actual content may be a film. There has to be a way to differentiate XXX and its content. Therefore, we need additional constructs such as "content in XXX." That is, the query may be of the following form.

SELECT FROM TABLE

WHERE CONTENT = IN XXX

We also need constructs to play-together, play-after and play-before. That is, to play video and audio data together, we may need a statement of the following form:

PLAY TOGETHER A AND B

WHERE A =

SELECT - - -

AND B =

SELECT - - -

That is, we need to play A and B together. To get A and B, we need select statements.

The above are just examples. A lot of work is in progress to define extensions to SQL and also to develop new multimedia query languages. Some examples are given in Proceedings of the IEEE Multimedia Database Systems Workshop[61,62] as well as in a special issue of *IEEE Transactions on Knowledge and Data Engineering*.[129] Note that SQL/MM contains the extensions that the standards groups have proposed for multimedia databases.

13.2.3 XML

XML was discussed in Part I of this book. It is one of the significant developments in information technology for the 1990s. What exactly is XML? Is it a data model, a metadata model, or something else completely? While different views have been given about XML, it can be viewed as everything. Essentially, it specifies a format that can be used to represent documents that can be universally understood. These documents could be text or multimedia documents, documents with relational data, or documents with financial data. Finally, we have some way to specify features in a common way, and, because the Web has millions of users, we need this for document representations.

Next, let us get into some specifics about XML. XML is a specification by the World Wide Web Consortium for document representations. Initially, it was developed to represent text documents. Text documents could be memos, letters, and papers. As stated by Rosenthal,[104] XML is a semistructured format for data with interesting tags. Tags are defined by tagsets called domain type definitions (DTDs). DTDs can be used to specify memos, letters, and other documents. XML is used only for specification. Its counterpart, XSL (extensible style language), is used for presenting a document. There are various application programming interfaces (APIs) for accessing XML content. Links between documents are provided by Xlink, which is a form of hyperlinking. Xpointer is used to point to something within a XML document.

XML evolved from HTML and SGML (standard generalized markup language). SGML was developed before the Web's time and involved too many details that were not necessary. HTML was developed for the Web and had limitations. For example, HTML has a fixed set of markup tags, and these tags do not help to understand the content. They are designed to help a browser correctly display the document. Consequently, the best search engines can index HTML documents based upon criteria like frequency of words. HTML cannot do one-to-many linking. Furthermore, it cannot extract pieces of text out of a document, and it cannot link to arbitrary portions of Web pages. These are just some of the deficiencies of HTML. XML attempts to overcome these deficiencies.[104]

XML provides the facility for creating one's own set of markup tags. That is, a document can be defined any way you want it. As long as the receiver's machine can understand XML tags, then the receiver can look at the document the way it was intended. Think of XML as a metalanguage. That is, XML is a language describing how to create one's markup language. By changing the tags, an XML document can get a completely different shape. As mentioned earlier, XSL is used for creating one's own set of presentation rules. Xlink, the XML link language, enables one-to-many linking and also bidirectional linking. Xpointer, the XML pointer language, enables pointing to something specific in a document without putting any anchor tags into the document. XML, XSL, XLink, and XPointer are the essential components for document representation for the Web. This is illustrated in Figure 13.2.

As mentioned earlier, various groups are proposing XML for representing documents such as financial securities, chemical structures, e-commerce product information, and multimedia data. One specific area of interest to the data management community is XML-QL, which is a query language for XML. As stated by Deutch et al.,[39] XML-QL is a declarative and relationally complete query language. Various proposals have been submitted to the World Wide Web Consortium for XML-QL. A simple XML-QL query extracts data from an XML document. For example, a query could be to extract the author and title from an XML document. A more complex query can perform joins on contents in XML documents as well as other complex operations. Queries can also be nested. An XML-QL has a data model, which is usually a graphical model, associated with it. A thorough discussion of one such XML-QL is given by Deutch.[39] Note that once the standard is adopted by the World Wide Web Consortium, there will be a common XML-QL.

One of the current limitations of XML is its inability to specify semantics. Some argue that it is not up to XML to specify semantics. Others argue that ontology work has to be integrated into XML. Resource descriptive format (RDF) has been developed specifically to handle this limitation of XML.[144] We may see some resolution in the next few years. Ontology, which is an important aspect of metadata, is the subject of the next section. It should be noted that XML specifications are continually evolving, like many standards. Therefore, we urge the reader to keep track of the developments (visit www.w3c.org). It should be noted that XML implementations may not conform entirely to the standards. So, users need to be aware of such issues before using an XML product. Also, extensions to handle multimedia data have also been proposed, and that language is called synchronized markup language (SMIL).[109]

13.2.4 ONTOLOGIES

We have heard a lot about ontologies in recent years. That is, terms have evolved from data dictionary to metadata and now to ontology. What exactly is an ontology? Fikes defined an ontology to be a specification of concepts used for expressing knowledge. This would include entities, attributes, relationships, and constraints.[96,145]

One may argue that we have been talking about entities and relationships for over two decades. So what additional benefits do ontologies give us? Onotologies are essentially an agreed-upon way to specify knowledge. Fikes states that ontologies are distinguished not by their form but by the role they play in representing knowledge. One can have ontologies to represent people, vehicles, animals, and other general entities such as tables, chairs, or chemistry. For example, a group of people could define an ontology for a person, and that ontology could be reused by someone else. Another group may want to modify the ontology for a person to have its own ontology for that person. That is, different groups could have different ontologies for the same entity. Once these ontologies are used repeatedly, a standard set of ontologies may evolve. There are efforts to standardize ontologies by different programs. In addition, standards organizations are also attempting to specify ontologies.

Why are ontologies useful? They are needed whenever two or more people have to work together. For example, ontologies are very important for collaboration, agent-to-agent communication, knowledge management, and for different database systems to interoperate. Ontologies are also useful for education and training, genetics, and modeling and simulation. In summary, many fields require ontologies. A good example is different groups collaborating on a design project. They should define ontologies so that they all speak the same language. If ontologies are previously defined by other design groups, these could be reused to save time.

We often hear about domain specific ontologies. What are they? One can arbitrarily come up with ontologies for aircraft. Groups working in various airforce organizations, however, may have their own specialization for aircraft. These are domain-specific ontologies. One challenge when interoperating heterogeneous databases is whether one can come up with a common set of ontologies for semantic integration of the databases or does he need to take each pair of databases and treat them separately? In order to come up with a common set of ontologies, it is sometimes necessary to examine various pairs, develop ontologies for these pairs, and then see if a common set of ontologies can be extracted. The goal for integrating heterogeneous databases is to come up with a common set of ontologies from the domain-specific ontologies. This is illustrated in Figure 13.3.

E-commerce applications and Web data management will define a new set of ontologies. For e-commerce applications, ontologies will include specifications for Web pages to set up e-commerce sites as well as ontologies for specifying various

Problem: Interoperation of Databases A, B, and C

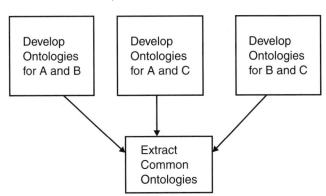

FIGURE 13.3 Common ontologies from domain-specific ontologies.

goods. There are Web sites emerging that specify ontologies. These ontologies can be used for various activities such as collaboration and integration.

A common question concerns the difference between ontologies and XML. While XML specifies the structure of a document, ontologies specify semantics of various applications. The challenge is to integrate the structure with the semantics to provide a complete set of interoperability mechanisms.

There is extensive research on ontologies by not only computer scientists, but also by logicians and philosophers. Uncertain reasoning, probabilities, and other heuristics are being incorporated into ontology research. As in the case of XML, this is also a very dynamic area, and the reader is urged to visit various Web sites specifying ontologies to keep up with the developments.

13.2.5 STORAGE STANDARDS

Storage standards in general and compression standards in particular have received a lot of attention for multimedia databases. As stated in Part I, multimedia data, especially images, voice, and video, can be quite large. Therefore, we need compression standards so that the compressed data use less space.

The compression standards for images include the JPEG standards.[73] Some compression standards may involve some loss of data, whereas others involve little or no loss of data. For video data, the compression standards include the MPEG (moving picture expert group) standards.[90] These standards have evolved over the past several years. Often, we have numbers attached to the standards such as MPEG 7, which specifies the particular version of the standards.

There are also markup standards. XML is an example. It is a document markup standard to exchange documents on the Web. It evolved from document markup languages such as SGML. Variations of markup languages for images, audio, and video have also evolved. For a discussion of the various standards, refer to various Web sites, such as *www.jpeg.org*[73] and *www.cselt.it/mpeg*.[90]

13.2.6 DATA MINING STANDARDS

Since data mining is now becoming a mature technology, it is important that appropriate standards be established for various aspects of data mining. For example, data mining processes have been developed, but these processes are yet to be standardized. We need to examine whether the various process models can be applied to model the data mining process. One group has developed various languages for data mining. For example, SQL extensions are being proposed. However, these extensions are yet to be made standard for data mining. Architecture for data mining is also being examined. We need to determine whether the various standards emerging from consortiums may be applied for data mining. Finally, data mining is becoming a key technology for e-business. The various standards for e-business need to be examined for data mining. In summary, as we make more and more progress in data mining, we cannot avoid standardization. Standardization will enable standard methods and procedures to be developed for data mining so that the entire process of data mining can be made easier for different types of users.

Clifton and Thuraisingham[27] address how standards may be applied for data mining. In particular, standards for the data mining process described in Part II are discussed.[30,101] Furthermore, SQL and data mining are also investigated. Next, architectures and standards for data mining were reviewed. The various standards and frameworks developed for e-business based on XML and data mining are also discussed (see also the discussion about frameworks such as Biztalk[65]).

13.2.7 MIDDLEWARE STANDARDS

One notable middleware standard is the object request broker, which is essentially a distributed object management (DOM) system. An example of DOM system that is being used as a middleware to connect heterogeneous database systems is a system based on Object Management Group's (OMG's) CORBA.* CORBA is a specification that enables heterogeneous applications, systems, and databases to interoperate with each other. There are three major components to CORBA.[95] One is the object model, which essentially includes most of the constructs discussed in Appendix B, the second is the object request broker (ORB), through which clients and servers communicate with each other, and the third is the interface definition language (IDL), which specifies the interfaces for client server communication. Figure 13.4 illustrates client–server communication through an ORB. In that figure, the clients and servers are encapsulated as objects. The two objects then communicate with each other through the ORB. Furthermore, the interfaces must conform to IDL. OMG is proposing extensions to IDL for multimedia and e-commerce data. For further details, refer to Orfali et al.[95,98]

13.2.8 OTHER STANDARDS

The previous sections discussed only some of the standards such as query language, XML, ontologies, storage, and middleware. Standardization efforts are underway in

---

* Note that middleware is referred to as the intermediate layer between the operating systems and the applications. This layer connects different systems and applications.

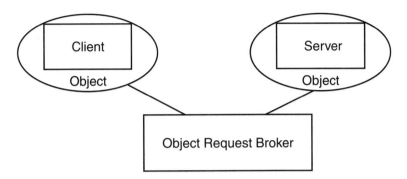

FIGURE 13.4 Interoperability through the ORB.

other areas, too. Some of these standards are national and international organizations such as ANSI (American National Standards Institute) and ISO (International Standards Organization), and others are consortia like OMG and some of the new ones for e-commerce frameworks. The World Wide Web Consortium is a consortium whose standards are being followed and adopted by many.

In addition to the standards mentioned here, there are also efforts to create standards for multimedia data modeling, architectures, and component-based database management. As we make more progress with research and development, we will see more development with standards. The next two sections discuss prototypes and products for multimedia data management and mining. These developments will contribute in return to standards, while the standards themselves influence the prototypes and products.

13.3 PROTOTYPES AND PRODUCTS FOR MULTIMEDIA DATA MANAGEMENT

13.3.1 OVERVIEW

Throughout Part I, we have discussed the status and challenges for MM-DBMSs from architectural, data modeling, and functional viewpoints. Much research has been carried out in recent years regarding MM-DBMSs. As a result, major database system vendors have now developed MM-DBMS products. However, many of these products use limited data management capabilities for multimedia databases. That is, a loose integration approach between a DBMS and a multimedia file manager is utilized. Many efforts have been reported on data modeling research for MM-DBMSs, and this field is fairly mature. Various extensions to object and relational models have been proposed to represent multimedia databases and also the relationships between data. Finally, various techniques for query, browsing, and filtering for multimedia data have been developed.

Current challenges in multimedia database management include developing appropriate storage management techniques, transaction management for multimedia databases, security, and integrity, as well as metadata management. While various access methods and indexing techniques such as content-based indexing have been

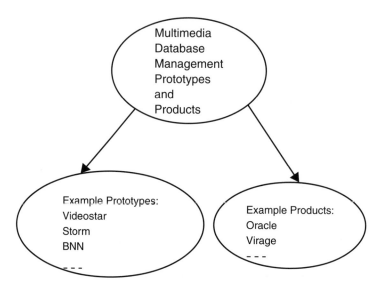

FIGURE 13.5 Multimedia database management prototypes and products.

developed, there is still a lot of work to be done to handle large video and audio databases. Transaction management research for multimedia databases is just beginning. Security and integrity issues have received little attention. Handling large metadatabases for multimedia databases is still an issue. Integrating real-time data management techniques with multimedia database management is an active research area. Finally, multimedia data management is also a challenge.

This section provides an overview of some of the prototypes and products. We first discuss the various prototypes, including Storm, VideoStar, and Broadcast News Navigation (BNN). We then discuss commercial products including Oracle Media Server and Virage. Figure 13.5 illustrates the prototypes and products discussed here.

13.3.2 PROTOTYPES

Various research prototypes and commercial products have been developed for multimedia database systems. Some of the recent prototypes include the storm multimedia database system developed in France[7] and the VideoStar system developed in Norway.[55] While there are many similarities between these two systems, there are also some major differences.

Both Storm and VideoStar systems manage multimedia data and provide support for querying as well as presenting the data. However, Storm uses an object-oriented database system to store the multimedia data and builds layers on top of the database system for various functions. The layer immediately above the database system models temporal relationships between the multimedia data. It is this layer that provides support to queries such as play-forward and play-backward. The layer on top of this modeling layer is the user interface that enables the presentation of multimedia data. VideoStar has no separate database system. That is, the data model captures not only the objects, but also the various relationships, including the temporal relationships between the objects. The database system manages these objects, i.e., there is no separate layer hosted on top of the database system to capture the temporal relationships. In addition to the two systems discussed here, there are many more systems that have been described in the literature.[92] A discussion of all of these systems is beyond the scope of this book.

BNN is a prototype developed at MITRE by Merlino and Maybury.[83] Essentially, it is a system which gathers video data, extracts stories from the data, uses Oracle's video server product (to be explained in the next subsesction) to store the summaries, and then uses the navigator to retrieve the summaries. So, for example, one could run CNN all night and let BNN produce the summaries. Then, the next day, one could retrieve relevant stories of interest. While BNN is not strictly a database management system and it uses a commercial database system, it does indexing and annotating to produce a complete multimedia application and data management system.

13.3.3 PRODUCTS

Numerous multimedia data management systems have been developed during recent years. Two such systems are discussed here: one is Oracle's video product and the other is a product by Virage.

Oracle Video Server is based on Oracle Video Cartridge.[97] As stated by Oracle, this technology enables corporations to write and deploy innovative multimedia business applications in their client/server environments. Oracle Video Cartridge expands Oracle's capabilities to support multimedia data management capabilities. In particular, Oracle Video Cartridge delivers the following:

- Simultaneous user support for full-motion, full-screen video and CD-quality audio at varied bit-rates
- Video annotation capability, which provides video indexing for fast access to specific content
- Integration with the Oracle database for a complete, enterprise-wide information management solution
- Application development support for multimedia information

Virage focuses entirely on video products. The Virage Internet Video Application Platform includes a suite of software solutions critical to organizations that create, manage, and distribute media on an enterprise scale.[134] Virage has a number of products, including their Video Application Server for managing multimedia data on Web sites and other products such as Videologger and Media analysis software. A number of white papers on Virage products can be found on www.virage.com.[134]

We have discussed only two products, one from a larger database corporation and another from a video server corporation. There are many more products from corporations such as Informix and Sybase. A discussion of all of these products is beyond the scope of this book.

13.4 PROTOTYPES AND PRODUCTS FOR MULTIMEDIA DATA MINING

13.4.1 OVERVIEW

This chapter describes some example commercial multimedia data mining products and research prototypes, some of which have evolved into products. As stated earlier, all of the information on these products has been obtained from published material as well as vendor product literature. Since commercial technology is advancing rapidly, the status of these products as described here may not be current. Again, our purpose is to give just an overview of what is out there and not the technical details of these products.

We discuss only some of the key features of the commercial products and prototypes. Note that various data management texts and conferences including data management/mining magazines, books, and trade shows such as *Database Programming and Design* (Miller Freeman Publishers), *Data Management Handbook Series* (Auerbach Publications), and DCI's Database Client Server Computing conferences contain several articles and presentations discussing commercial products. We urge the reader to take advantage of the information presented in these magazines, books, and conferences and to keep up with the latest developments with vendor products. Furthermore, in areas like World Wide Web mining, we can expect the developments to be changing very rapidly. The various Web pages are also a useful source of information.

We are not endorsing any of these products or prototypes. We have chosen a particular product or prototype only to explain a specific technology. We would have liked to have included discussions of many more products and prototypes, but such a discussion is beyond the scope of this book. In recent years, various documents have provided a detailed survey of data mining products and prototypes. One example is the document on data mining products by the Two Crows Corporation. This corporation periodically puts out detailed surveys of the products, and we encourage the reader to take advantage of such up-to-date information. There are also tutorials on comparing the various products.[76] Figure 13.6 illustrates the prototypes and products discussed here. Section 13.4.2 describes some of the prototypes, and Section 13.4.3 describes some of the products. Note that while many of these systems work for structured data, they are being adapted for unstructured data.

13.4.2 PROTOTYPES

Over the past three years, prototype data mining tools have emerged from various universities and research laboratories. Some of these tools are now commercially available. A discussion of all of these tools and systems is beyond the scope of this book, therefore, we have discussed only those tools that we are familiar with. Our discussion has been influenced by the work of Grupe and Owrang[49] (see also Thuraisingham[41]).

There are tools that attempt to develop new models for data mining. In particular, they work on relational as well as text data and provide frameworks for data mining. They are developed by the projects involved in developing new functional models. Example projects include one by Stanford University, the MITRE Corporation, and

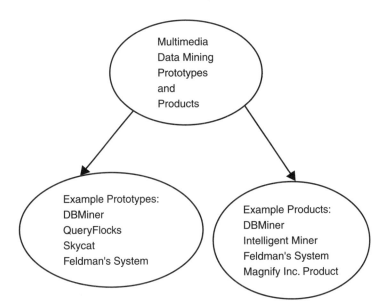

FIGURE 13.6 Multimedia data mining prototypes and products.

the Hitachi Corporation, and another by Rutgers University. The projects that attempt to develop new functional models essentially integrate data mining and database management. In particular, the tight integration discussed in Part II are taken by these initiatives. The project by Stanford University, the MITRE Corporation, and the Hitachi Corporation is called Queryflocks.[131] The idea here is to develop a query methodology and optimization techniques to handle flocks of queries to support data mining. In particular, these queries attempt to produce associations between entities in the database.

The project by Rutgers University also integrates data mining with database management and has formulated query languages for data mining queries. This project also attempts to find associations between the entities.[5] The projects that work on new information services essentially mine different types of data such as multimedia data. That is, multimedia data mining is the focus of the projects that fall under this category. We discuss some of them here.

Data mining on text is being attempted by the Queryflocks project by Stanford University, the MITRE Corporation, and the Hitachi Corporation. The technique for finding associations can be applied not only to relational databases but also to text databases once the tagged entities are extracted from the text. Another example of text mining is the work of Cheng and Ng at the University of Arizona. This project searches documents based on colocation of terms in documents. A third example is the work by Feldman at Bar-Ilan University in Israel, which finds association rules between identified concepts in text. Note that this latter tool is also now available as a commercial product.[5,131]

There is some work being done on image mining. Most notable is the SKICAT project being done at JPL (Jet Propulsion Laboratory). This work detects unusual objects from images in space.[5] Another example of image mining work is that of Clifton at the MITRE Corporation.[24] This work finds unusual patterns from hyper-spectral images. There is also work on image mining being done at the University of British Columbia.[91] The technique used therein is distance-based reasoning.

Some work on Web mining has also been reported. Examples include projects at the University of Michigan and the University of Minnesota.[58] In addition, there have been several new efforts reported regarding Web mining.[138] With respect to video mining, MITRE has applied Queryflocks to the output BNN system. Essentially, this is about video mining. BNN outputs stories, and Queryflocks mines those stories to find correlations.

Scalability of the algorithms in data mining is still a largely unexplored area. The Massive Digital Data Systems Project has focused on scalability of various data management and data mining techniques to handle very large databases.[82] Scalability could be determined by using larger and larger data sets or by conducting theoretical as well as simulation studies.

Scalability of data mining algorithms needs a lot of work. There is some work being done at Magnify Inc. to determine the scalability of specific data mining techniques using, among other things, image data. These algorithms handle terabytes of data. Other products focusing on scalability of data mining techniques include those by the Thinking Machines Corporation and SGI (Silicon Graphics). The work being performed at IBM's Yorktown Heights research laboratory also addresses scalability issues.

Part II discussed the data mining process. After cleaning and mining the data, one needs to extract only the useful information. Therefore, understanding the data becomes very important. Some research projects focus on this aspect of data mining. We discuss a few here.

GTE (General Telecommunications and Equipment) Laboratories has worked on data mining for a number of years. They focus on understanding the data by producing domain-specific reports. Medical cost mining is an application area for this work. Another effort is being carried out at Simon Fraser University, one of the prominent research locations for data mining. One aspect of this work is integrating with visualization tools so that the data mining results can be better understood.[52] A third effort is the work being carried out at the University of Massachusetts at Lowell.[48] That work also focuses on integrating data mining with visualization techniques. While these efforts focus mainly on relational data, we feel that understanding the results is very important for multimedia data mining.

At present, much of the focus has been on applying data mining tools to extract patterns. Many say that this is what data mining is all about, that understanding the results is the responsibility of some other area. However, if data mining is to be useful, we need to focus on understanding as well as mining the data.

Two very prominent projects in data mining include IBM's Quest project, by Agarwal et al., and Simon Fraser University's DBMiner product, by Han et al. Numerous papers have been published on this work.[108] We discuss only the essential points.[5]

IBM's Quest project uses multiple data mining techniques and finds sequential associations as well as time-series associations. The work is influenced by database systems technology and builds data mining techniques to work with relational database systems. Some of this research has been transferred to IBM's products (discussed in Section 9.3). Simon Fraser University's DBMiner is now available as a product that focuses on mining relational data that has been warehoused and includes end-user support, visualization capabilities, and the understanding of results. Mining association rules is the major focus of both the efforts. While these projects initially focused on relational data, more recently they have focused on multimedia data.

The two large-scale projects that we have discussed here have influenced several other efforts that have emerged. Describing all of these efforts is beyond the scope of this book. However, recent conferences such as the ACM SIGMOD conference, the IEEE Data Engineering conference, the VLDB conference, as well as various data mining conferences listed earlier contain many research papers that describe other emerging large-scale projects.[9,57,74,100,108,137] Additional research papers on data mining, such as that by Fayyad,[42] can also be found. Up-to-date information on products can be found on the KDD Nuggets Web site.[75]

13.4.3 PRODUCTS

Many data mining commercial tools are now emerging. These include IBM's Intelligent Miner, Information Discovery's IDIS, and Neovista's Decision Series. While many of these work on relational data, they are being adapted to handle nonrelational data. This section provides an overview of some of the tools.

Other data mining products include Whizsoft's Whizwhy, which is an end-user association rule-finding tool that uses rule-based reasoning. Hugin's product, also called Hugin, uses Bayesian reasoning and is good for prediction. Data Logic/R, the product of Reduct Systems, uses rough sets as a data mining technique, and Nicel, by Nicesoft, uses fuzzy logic as a data mining technique. SGI's Mineset integrates data mining with visualization and focuses on high performance data mining. The product Darwin by Thinking Machines Corporation also illustrates high performance data mining. SRA Corporation's product finds patterns for fraud detection. MRJ Corporation's data mining product does mining on large data sets. Other notable products include the SAS Institute's Enterprise Miner and Redbrick's Datamind. There are also many more products on the market, and, as mentioned earlier in this chapter, describing all of them is beyond the scope of this book. A good tutorial comparing the various products was presented at the Knowledge Discovery in Databases Conference in 1998.[76]

IBM's Intelligent Miner is a popular data mining product on the market. This product incorporates some of the research that has come out of IBM's Almaden Research Center. In particular, the Quest research project at Almaden has produced some of the origins of the Intelligent Miner product.

The techniques used by this product are many. The Intelligent Miner is, in a way, a multi-strategy data miner. In particular, it uses association rules, decision trees, neural networks, and nearest neighbor methods for mining. It selects the methods as appropriate for the particular task. The outcomes it handles include missing values, anomaly detection, and categorization of continuous data, and it has applications in various domains. One of the conditions for this product is that the data has to reside in IBM's database system product called DB2. However, this may change with time. The product is available for PC as well as mainframe environments. There is work being done to adapt this tool for text data.

Information Discovery's IDIS is one of the earlier data mining tools running on Microsoft's Windows as well as NT environments, and it is an end-user information discovery tool. It provides natural language reports and uses various induction and machine learning techniques. It operates on smaller relational databases. Producing natural language reports of the data mining activities will help to understand the data mining results. Essentially, IDIS hypothesizes rules and then tests to see if the hypothesis is valid. It outputs unusual patterns as well as patterns that deviate from the norm. IDIS has been used for a variety of applications in fields such as financial, medical, scientific research, and marketing.

Neo Vista's Decision Series product attempts to provide a framework for integrating multiple data mining products. It supports several data mining techniques such as association rules, neural networks, nearest neighbor algorithms, and genetic algorithms. This framework has tight integration with open database connectivity (ODBC) on the server side. ODBC provides the glue for integrating the various data mining products. The data miners can access data from multiple relational database systems, which are integrated via ODBC. Neo Vista's product essentially focuses on middleware for data mining. This is one of the first products to provide such a capability, which is necessary if multiple data mining strategies and products are to be integrated to develop more sophisticated data mining tools.

13.5 SUMMARY

This chapter began with a discussion of standards for multimedia data management and mining. In particular, we discussed query language, XML, ontologies, storage, and mining standards. Then we provided an overview of multimedia data management prototypes and products, and, finally, we discussed multimedia data mining prototypes and products. As mentioned earlier, we are not endorsing any of the products. Furthermore, due to the rapid developments in the field, the information about these products may soon be outdated. Therefore, we urge the reader to take advantage of the various commercial and research material available on these products.

The developments in data mining in general and multimedia data mining in particular over the last few years have shown a lot of promise. Although some of the products have been around for a while, they are now being integrated with databases and information retrieval systems. As mentioned previously, we need the integration of multiple technologies to make data mining work. Furthermore, having good data is critical. Therefore, in the future, we will see more and more mining tools being integrated with various types of database systems as well as warehouses.

Conclusion to Part III

Part III dealt with multimedia on the Web. Chapter 10 provided an introduction to multimedia database management and mining on the Web. Chapter 11 examined the applications of multimedia data management and mining for collaboration, knowledge management, and training. Chapter 12 described security and privacy aspects. Finally, standards, prototypes, and products for multimedia data management and mining were discussed in Chapter 13.

Since multimedia data management in general and multimedia data mining in particular are still new technologies, there is still much to be done. This book has discussed some of the developments and challenges. The area is advancing so rapidly that information that is current today may be outdated tomorrow. As we have stressed in this book, the reader is urged to keep up with the developments of the various prototypes and products. The Web is an excellent resource to obtain current information.

14 Summary and Directions

14.1 ABOUT THIS CHAPTER

This chapter brings us to the end of this book. We have discussed various multimedia data management technologies and issues of multimedia data mining. We described how multimedia might be applied to the Web and electronic commerce. As stated in this book, multimedia for the Web is an integration of multiple technologies and has applications in e-commerce. This chapter summarizes the contents of this book and then provides an overview of the challenges and directions in multimedia data management and mining for the Web. We also give the reader some suggestions as to where to go from here.

The organization of this chapter is as follows. Section 14.2 summarizes the contents of this book. Note that each of the previous chapters (2–13) included a summary at the end of the chapter. Essentially, we have collected these summaries and put them together to form an overall summary of the book. Then, Section 14.3 discusses the challenges and directions for multimedia data management. Section 14.4 discusses the challenges and directions for multimedia data mining. Section 14.5 discusses challenges and directions for multimedia on the Web. Essentially, we consider the trends discussed in this book, and, for each of the topics addressed, we have discussed future work in the area. One could argue that the directions are also part of the challenges. In addressing the directions, one needs to address the challenges as well. Finally, Section 14.5 also includes suggestions to the reader as to where to go from here. Some of the key points in this book are reiterated, and the reader is encouraged to take the steps to make multimedia data management and mining a success.

14.2 SUMMARY OF THIS BOOK

Figure 14.1 duplicates Figure 1.6 to recap what has been described throughout this book. Chapter 1 provided an introduction to multimedia data management, data mining, and multimedia for e-commerce. We also discussed the three parts of this book. Part I, describing multimedia data management, consists of Chapters 2, 3, 4, 5, 6, and 7. Part II, describing multimedia data mining, consists of Chapters 8 and 9. Part III, describing multimedia for the Web, consists of Chapters 10, 11, 12, and 13. Chapters 2 through 13 are summarized in this section.

Chapter 2 discussed various types of architectures for multimedia database systems. In particular, loose/tight integration, schema, system, functional, distributed, interoperability, and hypermedia architectures were discussed. Essentially, we examined architectures for database systems and discussed the impact of multimedia data management on these architectures. In the case of interoperability architectures, we

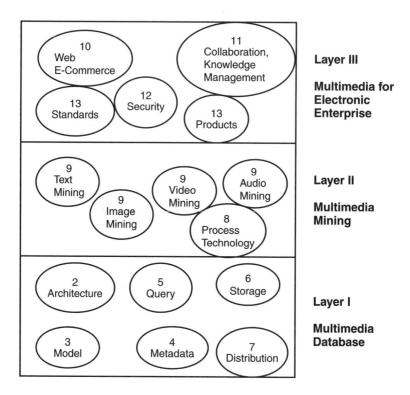

FIGURE 14.1 Topics addressed in this book.

examined various aspects, including architectures for integrated heterogeneous databases, architectures based on object request brokers, client-server-based architectures, three-tier architectures, and component-based architectures. We believe that component-based architectures will gain popularity as more and more applications deal with complex data such as voice, text, video, and images.

Chapter 3 discussed various aspects of data and information modeling for multimedia databases and applications. First, we provided an overview of data models, including object-oriented data models, object-relational data models, and hypersemantic data models. We then discussed information modeling.

Chapter 4 described metadata for multimedia data. We started with a discussion of various types of metadata. In particular, metadata for text, images, audio, and video were described. Both content-dependent as well as content-independent metadata were discussed. We also discussed annotations for multimedia data. In addition, we dealth with issues of extracting metadata from the data for text, video, audio, and images. Next, we addressed various aspects of metadata, five topics in particular. One deals with metadata for combinations of data types, the second deals with ontologies, the third deals with annotations, the fourth deals with pedigree, and the fifth deals with XML for multimedia data. As mentioned in this chapter, XML is becoming one of the important developments for specifying multimedia documents.

Finally, we discussed issues of managing the metadata. In particular, issues of querying the metadata, carrying out transactions on metadata, as well as distributing the metadata were discussed. Often, for multimedia databases, the metadata is quite large. Therefore, efficient techniques are needed for managing the metadata.

Chapter 5 provided an overview of multimedia query processing. First, we discussed data manipulation issues, in particular, browsing, editing, transaction processing, querying, and updating. We then discussed query processing at some length. We gave examples of performing joins in multimedia databases. Then we discussed query language issues for multimedia database systems.

Chapter 6 briefly described access methods and indexing and storage strategies for multimedia databases. Some key issues for efficient query processing were addressed in this chapter. We started with a discussion of indexing for multimedia databases and gave some examples. We can index data with keywords, or we can use multimedia data such as images for indexing. Next, we discussed caching as well as synchronizing the display of multimedia data. Finally, we discussed storage strategies for multimedia data, including single disk storage as well as multiple disk storage; we illustrated the discussion with examples.

Chapter 7 described the impact of multimedia data on distributed database systems. That is, we discussed various aspects of distributed multimedia database systems. For example, one node may have audio data and another may have video data. In some cases, the multimedia database may also be distributed. Chapter 7 began with a discussion of distributed architectures and database design. We then focused on major functions such as distributed query processing, transaction management, metadata management, and security and integrity management. We discussed interoperability, migration, and warehousing aspects for multimedia database management. As mentioned, much of the work on warehousing and mining have been carried out on relational databases. However, as we get more and more multimedia data on the Web, it is important to be able to warehouse the data and then apply the mining tools. Therefore, multimedia data warehouses will become popular in the future.

Chapter 8 took various data mining prerequisites and examined their impact on multimedia data. We started with a discussion of multimedia data mining technologies. These include multimedia database management and multimedia warehousing, as well some traditional data mining technologies such as statistical reasoning and visualization. We also provided an overview of architectural support. We then discussed the need for data mining. We gave several examples. Finally, an overview of data mining techniques was described. Many of these techniques may be applied for multimedia data mining, too.

Chapter 9 mainly addressed mining individual data types. Mining multimedia data involves addressing a combination of two or more media types. As we learn more about mining text, images, video, and audio, we can expect progress to be made in multimedia data mining. Mining combinations of data such as video and text, video, audio, and text, or image and text remains a challenge.

Chapter 10 discussed the applications of multimedia data management and mining for the electronic enterprise. In particular, the applications of multimedia technology for the Web and e-commerce were described. First, we started with a discussion of multimedia data processing issues in general. We then focused on both multimedia data management as well as multimedia data mining aspects for the Web and e-commerce. We also rounded out the key topics with discussions of agents for multimedia data mining, mining distributed multimedia databases, and metadata mining.

Chapter 11 discussed the applications of multimedia data management for key information technologies for the Web. These include collaboration, knowledge management, and training. We also discussed some other technologies such as real-time computing, visualization, and quality of service aspects.

Chapter 12 was devoted to the important area of security and privacy as related to the Web as well as privacy issues for data mining. While there have been efforts to apply data mining for handling security problems such as intrusion detection, this chapter focused on the negative effects of data mining. In particular, we discussed the inference problem that can result due to mining as well as ways of compromising privacy because of Web data access. Finally, we discussed some general security measures for the Web. Next, we focused on multimedia considerations since the focus of this book is on managing and mining multimedia databases. All of the security and privacy concerns apply for multimedia as well as non-multimedia databases. However, with multimedia data we have additional considerations such as controlling access to parts of images and video data. We also briefly addressed data quality and integrity issues for multimedia databases.

Finally, Chapter 13 began with a discussion of standards for multimedia data management and mining. In particular, we discussed query language, XML, ontologies, storage, and mining standards. We then provided an overview of multimedia data management prototypes and products, and, finally, we provided a discussion of multimedia data mining prototypes and products. As mentioned earlier, we are not endorsing any of the products. Furthermore, due to the rapid developments in the field, the information about these products may soon be outdated. Therefore, we urge the reader to take advantage of the various commercial and research material available on these products.

14.3 CHALLENGES AND DIRECTIONS FOR MULTIMEDIA DATABASE MANAGEMENT

Challenges in multimedia data management include developing architectures, data models, query strategies, storage management, metadata management, distribution, security, and integrity. Part I discussed various multimedia data management functions. We need to develop systems that can function on the Web. For example, efficient query optimization strategies for the Web are needed. We also need to develop techniques for presenting multimedia data on the Web. We need to synchronize storage management with presentation. Below, we summarize the directions for multimedia data management.

- Novel architectures for managing multimedia data
- Data models for multimedia data
- Efficient query processing strategies
- Transaction models for multimedia data
- Storage management for multimedia data

- Data warehouses for multimedia data
- Integrating structured data with unstructured data
- Data distribution issues for multimedia data
- Secure multimedia data management strategies
- Query languages for multimedia data
- Metadata management for multimedia data
- XML extensions for multimedia data
- Integrating heterogeneous multimedia databases
- Data integrity for multimedia data
- QoS for multimedia database management
- Real-time scheduling for multimedia data
- Techniques for video on demand
- Indexing and access methods for multimedia data
- Managing multimedia data on the Web

14.4 CHALLENGES AND DIRECTIONS FOR MULTIMEDIA DATA MINING

The challenges for multimedia data mining include mining individual data types as well as combinations of data types. First, we need to examine data mining technologies and techniques and determine the impact of mining multimedia data. We addressed some of the issues in this book, but there are numerous areas for further research and development. For example, what outcomes does one need to look for when mining video, image, and audio data? We discussed anomaly detection in this book. Are there other outcomes such as associations that may be suitable for audio, image, and video data? Does one need to convert multimedia data to relational data and then mine the relational data? Can one mine multimedia data directly? Below, we summarize the directions for mining multimedia data.

- Techniques for mining text, image, audio, and video as well as combinations of data types
- Examining data mining technologies and techniques and determining the impact of mining multimedia data
- Scalability of multimedia data mining
- Mining multimedia data on the Web
- Mining distributed multimedia data
- Metadata for multimedia data mining
- Security and privacy issues for data mining
- Architectures for multimedia data mining

14.5 CHALLENGES AND DIRECTIONS FOR MULTIMEDIA FOR THE WEB AND E-COMMERCE

Whether we are developing techniques for multimedia data management or mining, they have to work on the Web. That is, multimedia data management and mining on the Web are becoming a critical need. Therefore, we need to examine all of the issues and developments discussed in this book and examine the impact of the Web. Multimedia technology has applications in e-commerce. We discussed some of them, but there is lot of work yet to be done. Below, we summarize some of the directions for multimedia for the Web and e-commerce.

- Multimedia for knowledge management and collaboration
- Security and privacy issues
- Multimedia for e-commerce
- Multimedia data management techniques for the Web
- Multimedia data mining techniques for the Web

14.6 WHERE DO WE GO FROM HERE?

We have provided a broad overview of multimedia data management and mining and their applications to the Web and e-commerce. In particular, we discussed technologies, techniques, tools, and trends. We have also given many references should the reader need in-depth coverage of a particular topic. However, all the literature is not going to give the reader a better appreciation of what multimedia data management and data mining are all about. It is certainly useful to have a good knowledge of multimedia technologies and be able to speak intelligently about it. However, in order to know what technique works, the limitations of an algorithm, or how to conduct e-commerce, hands-on experience with the tools is necessary. As in the case of many technologies, multimedia data management and mining gets better with practice; we urge the reader to work with practical applications in using the multimedia data management tools as well as with developing the tools, if possible.

Starting data mining or an e-commerce project, especially with multimedia technologies, requires management buy-in. This is partly because many of the technologies discussed here are rather new, which means financial and personnel resources are needed. Furthermore, you need to decide whether to contract the work or have it done in-house. If you are using a commercial tool, then you need to have frequent communication with the developer. In other words, the customer, the e-commerce tool/solutions developer, the data mining tools/solutions developer, and those who carry out e-commerce and data mining have to work very closely together, otherwise the project may be a failure.

We believe that there are exciting opportunities in multimedia, data mining, data management, and e-commerce with the emergence of new Web data management technologies. Furthermore, technology integration, such as integration of data management, data mining, objects, and security, is making a lot of progress. As the user gets flooded with more and more multimedia data and information, the need to analyze this information, give only the information the user needs, and extract previously unknown information to help the user in the decision making process will become urgent. Also, various laws and policies will be clearer with e-commerce, and corporations as well as consumers will soon become familiar with the process of e-commerce. We can also expect universities to offer special courses and special-ization certificates on e-commerce, data mining, and multimedia technologies. We feel that the opportunities and challenges in multimedia technologies in general and e-commerce in particular will be endless.

Some of the key points discussed in this book are listed below. It should be noted that as we make progress, the issues will be much better understood.

- The Web will be the integration platform for all types of multimedia data, information, and knowledge management technologies.
- The various multimedia data, information, and knowledge management technologies have to work with the Web. This will include database access through the Web and collaboration on the Web.
- We need infrastructures for the Web to support the various data, information, and knowledge management technologies.
- E-commerce and e-business will thrive only if we successfully establish the Web as the integration platform.
- We need research and development programs to enable the Web to be the integration platform. This means that whenever we conduct research on data, information, and knowledge management technologies, we cannot ignore the Web.
- Data mining will continue to explode. We must develop techniques for mining text, audio, image, and video data. In addition techniques for mining combinations of data types are also critical.
- System integration is key. Multimedia database systems have to be inte-grated with data mining and applications systems. The resulting systems also have to be integrated with e-commerce systems.
- Standards such as XML, RDF, and related technologies will continue to play a major role in multimedia database products as well as e-commerce products.

References

1. Special Issue on Next Generation Database Systems, *Commun. ACM,* October 1991.
2. Special Issue on Computer Support Cooperative Work, *Commun. ACM,* November 1991.
3. Proceedings of the ACM Multimedia Database System Workshop, October 1994.
4. Working Notes on ACM Multimedia Database System Workshop, November 1995.
5. Special Issue on Data Mining, *Commun. ACM,* November 1996.
6. Proceedings of the ACM Multimedia Conference, 1994–1996.
7. Adiba, M., STORM: an object-oriented multimedia dbms, in *Multimedia Database Systems,* Nwosu, K. Thurainghsma, B., and Berra, B., Eds., Kluwer Publishers, Norwood, MA, 1996.
8. Adriaans, P. and Zantinge, D., *Data Mining,* Addison-Wesley, Reading, MA, 1996.
9. Proceedings of the First Federal Data Mining Symposium, Washington, D.C., December 1997.
10. Air Force Summer Study Board Report on Multilevel Secure Database Systems, Department of Defense Document, Washington, D.C., August 1983.
11. Proceedings of the Symposium on Advanced Information Processing and Analysis, Tysons Corner, March 1995.
12. Proceedings of the Symposium on Advanced Information Processing and Analysis, Tysons Corner, March 1996.
13. Alonso, G. et al., Exotica/FDMC: a workflow management system for mobile and disconnected clients, *Distributed Parallel Databases J.,* March 1996.
14. Banerjee, J. et al., A data model for object-oriented applications, *ACM Trans. Office Inf. Syst.,* October 1987.
15. Bell, D. and Grimson, J., *Distributed Database Systems,* Addison-Wesley, Reading, MA, 1992.
16. Benyon, D., *Information and Data Modeling,* Blackwell Scientific Publications, Oxford, UK, 1990.
17. Berry, M. and Linoff, G., *Data Mining Techniques,* John Wiley & Sons, New York, 1997.
18. Binn, L., Hypersemantic data modeling for secure data management, Proceedings of the IFIP Database Security Conference, Hildesheim, Germany, August 1994.
19. Carbone, P., Data mining, in *Handbook of Data Management,* Thuraisingham, B., Ed., Auerbach Publications, New York, 1998.
20. Ceri, S. and Pelagatti, G., *Distributed Databases, Principles and Systems,* McGraw-Hill, New York, 1984.
21. Chorafas, D., Intelligent Multimedia Databases, Prentice-Hall, Englewood Cliffs, NJ, 1994.
22. Clifton, C. and Marks, D., Privacy issues in data mining, Proceedings of the SIGMOD Data Mining Workshop, Montreal, 1996.
23. Clifton, C., Text mining, private communication, Bedford, MA, January 1996.
24. Clifton, C., Image mining, private communication, Bedford, MA, July 1998.
25. Clifton, C., Data mining and security, Proceedings of the IFIP Conference on Database Security, Seattle, WA, July 1999.

26. Clifton, C., Data mining for intrusion detection, IFIP Database Security Conference Panel, Seattle, WA, July 1999.

27. Clifton, C. and Thuraisingham, B., Standards for data mining, *Comput. Stand. Interfaces J.,* Accepted, 2001.

28. Codd, E. F., A relational model of data for large shared data banks, *Commun. ACM,* May 1970.

29. Cooley, R., Taxonomy for Web mining, private communication, Bedford, MA, August 1998.

30. Cross industry standard process for data mining, *http://www.crisp-dm.org.*

31. Workshop on Knowledge Discovery in Databases, Defense Advanced Research Projects Agency, Pittsburgh, PA, June 1998.

32. Date, C. J., *An Introduction to Database Management Systems,* Addison-Wesley, Reading, MA, 1990.

33. Davenport, T., *Working Knowledge: How Organizations Manage What They Know,* Harvard Business School Press, Boston, MA, 1997.

34. Decker, S. et al., The semantic Web: the roles of XML and RDF, *IEEE Internet Comput.,* October 2000.

35. *Decision Support Journal,* Elsevier/North Holland Publications, Amsterdam.

36. DeGroot, T., *Probability and Statistics,* Addison-Wesley, Reading, MA, 1986.

37. Workshop on Distributed and Parallel Data Mining, Melbourne, Australia, April 1998.

38. DSV Laboratory, Inductive logic programming, private communication, Stockholm, Sweden, June 1998.

39. Deutch, A. et al., XML-QL: a query language for XML, *http://w3c1.inria.fr/TR/1998/NOTE-xml-ql-19980819/.*

40. DiPippo, L., Hodys, E., and Thuraisingham, B., Towards a real-time agent architecture: a white paper, Proceedings of IEEE WORDS, Monterey, CA, November 1999.

41. Thuraisingham, B., Ed., *Data Management Handbook Supplement,* Auerbach Publications, New York, NY, 1998.

42. Fayyad, U. et al., *Advances in Knowledge Discovery and Data Mining,* MIT Press, Cambridge, MA, 1996.

43. Feldman, R. and Dagan, I., Knowledge discovery in textual databases (KDT), Proceedings of the 1995 Knowledge Discovery in Databases Conference, Montreal, Canada, August 1995.

44. Fowler, M. et al., *UML Distilled: Applying the Standard Object Modeling Language,* Addison-Wesley, Reading, MA, 1997.

45. Geppert, A. and Dittrich, K., Component database systems, in *Advances in Database Technology and Design,* Piattini, M. and Diaz, O., Eds., Artech House, Norwood, MA, 2000.

46. Ghosh, A., *E-commerce Security, Weak Links and Strong Defences,* John Wiley and Sons, New York, NY, 1998.

47. Grasso, A. et al., Data mining for e-business, Proceedings of the SPIE Data Mining Conference, Orlando, FL, April 2000.

48. Grinstein, G. and Thuraisingham, B., Data mining and visualization: a position paper, Proceedings of the Workshop on Databases in Visualization, Atlanta, GA, October 1995.

49. Grupe, F. and Owrang, M., Database mining tools, in *Handbook of Data Management Supplement,* Thuraisingham, B., Ed., Auerbach Publications, New York, NY, 1998.

50. Guo, Y., Kensington data mining, *http://ruby.doc.ic.ac.uk.*

51. Guttman, A., R-trees: a dynamic index structure for spatial searching, Proceedings of the ACM SIGMOD Conference, Boston, MA, 1984.

52. Han, J., Data mining, keynote address, Second Pacific Asia Conference on Data Mining, Melbourne, Australia, April 1998.

53. *Harvard Business School Articles on Knowledge Management,* Harvard University Press, Boston, MA, 1996.

54. Hinke T., Inference and aggregation detection in database management systems, Proceedings of the 1988 Conference on Security and Privacy, Oakland, CA, April 1988.

55. Hjelsvold, R. et al., Searching and browsing a shared video database, in *Multimedia Database Systems,* Nwosu, K., Thuraisingham, B., and Berra, B., Eds., Kluwer Publishers, Norwood, MA, 1996.

56. Hoschka, P., Synchronized Multimedia Integration Language, World Wide Web Consortium Report, June 1998.

57. Proceedings of the IEEE Data Engineering Conference, Orlando, FL, 1998.

58. Panel on Web mining, International Conference on Tools for Artificial Intelligence, Newport Beach, CA, November 1997.

59. Hurson, A. et al., Ed., Parallel Architectures for Databases, IEEE Tutorial, 1989.

60. Special Issue on Multidatabase Systems, *IEEE Comput,,* December 1991.

61. Proceedings of the IEEE Multimedia Database Systems Workshop, Blue Mountain Lake, N.Y., August 1995.

62. Proceedings of the IEEE Multimedia Database Systems Workshop, Blue Mountain Lake, N.Y., August 1996.

63. Proceedings of the IEEE Multimedia Database Systems Workshop, Dayton, OH, August 1998.

64. Special Issue on Collaborative Computing, *IEEE Comput.,* September 1993.

65. Special Issue on E-Commerce, *IEEE Comput.,* October 2000

66. Proceedings of the IEEE Multimedia Systems Conference Series, 1996–1998.

67. Proceedings of the IFIP Conference Series on Database Security, North Holland, 1998–1994.

68. Panel on data warehousing and data mining security, Proceedings of the IFIP Conference Series on Database Security, North Holland, Lake Tahoe, CA, August 1997.

69. Panel on data mining and Web security, Proceedings of the IFIP Conference Series on Database Security, North Holland, Thessalonicki, Greece, July 1998.

70. Summer school on inductive logic programming, Prague, Czech Republic, September 1998.

71. Inmon, W., *Building the Data Warehouse,* John Wiley and Sons, New York, NY, 1993.

72. Jensen, D., Real-time Java, Proceedings of the ISORC Symposium, Newport Beach, CA, March 2000.

73. *http://www.jpeg.org/public/jpeghomepage.htm.*

74. Proceedings of the Fourth Knowledge Discovery in Databases Conference, New York, NY, August 1998.

75. *http://www.kdnuggets.com.*

76. Elder, J. and Abbott, D., Tutorial on commercial data mining tools, Knowledge Discovery in Databases Conference, New York, August 1998.

77. Kim, W. et al., *Query Processing in Database Systems,* Springer-Verlag, New York, NY, 1985.

78. Korth, H. and Silberschatz, A., *Database System Concepts,* McGraw-Hill, New York, NY, 1986.

79. Lin, T.Y., Ed., *Rough Sets and Data Mining,* Kluwer Publishers, Norwood, MA, 1997.

80. Lloyd, J., *Logic Programming,* Springer-Verlag, Heidelberg, 1987.

81. Marshak, R., Workflow computing, *Patricia Seybold's Office Comput.,* March 1992.

82. *Proceedings of the Massive Digital Data Systems Workshop,* Community Management Staff, Washington D.C., 1994.

83. Merlino, A. et al., Broadcast news navigation using story segments, Proceedings of the 1997 ACM Multimedia Conference, Seattle, WA, November 1998.

84. Musick, R., *Proceedings of the First IEEE Metadata Conference,* Lawrence Livermore National Laboratory, Silver Spring, MD, April 1996.

85. Mitchell, T., *Machine Learning,* McGraw-Hill, New York, NY, 1997.

86. Morey, D., Knowledge management architecture, in *Handbook of Data Management,* Thuraisingham, B., Ed., Auerbach Publications, New York, NY, 1998.

87. Morey, D., Web mining, private communication, Bedford, MA, June 1998.

88. Morey, D., Maybury, M., and Thuraisingham, B., Eds., *Knowledge Management,* MIT Press, Cambridge, MA, 2001.

89. Morgenstern, M., Security and inference in multilevel database and knowledge base systems, Proceedings of the ACM SIGMOD Conference, San Francisco, CA, June 1987.

90. *http://www.cselt.it/mpeg/.*

91. Ng, R., Image mining, private communication, Vancouver, British Columbia, December 1997.

92. Nwosu, K. et al., Eds., *Multimedia Database Systems, Design and Implementation Strategies,* Kluwer Publishers, Norwood, MA, 1996.

93. *http://www.odi.com/excelon/main.htm.*

94. Oomoto, E. and Tanaka, K., OVID: design and implementation of a video object database system, *IEEE Trans. Knowledge Data Eng.,* April 1993.

95. *Common Object Request Broker Architecture and Specification,* OMG Publications, John Wiley and Sons, New York, NY, 1995.

96. *http://www-db.stanford.edu/LIC/HPKBtalk/sld002.htm.*

97. *http://technet.oracle.com/products/oracle7/htdocs/media/info/xovsds3. htm.*

98. Orfali, R. et al., *The Essential, Distributed Objects Survival Guide,* John Wiley and Sons, New York, NY, 1996.

99. Oszu, T. and Valdurez, P., *Distributed Database Systems,* Prentice-Hall, Englewood Cliffs, NJ, 1992.

100. Proceedings of the Knowledge Discovery in Databases Conference, New York, N.Y., August 1998.

101. PMML 1.0 — Predictive Model Markup Language, Data Mining Group, *http://www.dmg.org/public/techreports/pmml-1_0.html.*

102. Prabhakaran, B., *Multimedia Database Systems,* Kluwer Publishers, Norwood, MA, 1997.

103. Quinlan, R., *C4.5: Programs for Machine Learning,* Morgan Kaufmann, CA, 1993.

104. Rosenthal, A., private communication on XML, Bedford, MA, February 1999.

105. Rumbaugh, J. et al., *Object Modeling Technique,* Prentice-Hall, Englewood Cliffs, NJ, 1991.

106. Sheth, A. and Larson, J., Federated database systems, *ACM Comput. Surv.,* September 1990.

107. Sheth, A., and Rusinkiewicz, M., On transactional workflows, *IEEE Data Eng. Bull.,* March 1993.

108. Proceedings of the ACM SIGMOD Conference, Seatlle, WA, 1998.

109. *http://www.w3.org/TR/REC-smil/.*

110. SQL3, American National Standards Institute, draft, Washington, D.C., 1992.

111. *http://www.jcc.com/SQLPages/SQL%20Multimedia.htm.*

112. Thomsen, C. et al., Open OODB, *IEEE Comput.,* October 1992.

113. Thuraisingham, B., Security checking in relational database systems augmented by an inference engine, *Comput. Security,* December 1987.

114. Thuraisingham, B., Security for object-oriented database systems, Proceedings of the ACM OOPSLA Conference, New Orleans, LA, October 1989.

115. Thuraisingham, B., Recursion theoretic properties of the inference problem, MITRE Report, Bedford, MA, June 1990.

116. Thuraisingham, B., Nonmonotonic typed multilevel logic for multilevel secure database systems, MITRE Report, June 1990.

117. Thuraisingham, B., Novel approaches for the inference problem, Proceedings of the RADC Workshop on Database Security, Castile, N.Y., June 1990.

118. Thuraisingham, B., Multilevel security for multimedia database systems, Proceedings of the IFIP Database Security Conference, Halifax, UK, September 1990.

119. Thuraisingham, B., On the use of conceptual structures to handle the inference problem, Proceedings of the 1991 IFIP Database Security Conference, Shepherdstown, WV, November 1991.

120. Thuraisingham, B., Object approach for PACS system, *MITRE Inf. Syst. J.,* November 1993.

121. Thuraisingham, B. et al., Design and implementation of a database inference controller, *Data Knowledge Eng. J.,* December 1993.

122. Thuraisingham, B. and Lavender, B., Distributed multimedia databases, Proceedings of the APIA Symposium, Washington, D.C., March 1994.

123. Thuraisingham, B. and Ford, W., Security constraint processing in a multilevel secure distributed database management system, *IEEE Trans. Knowledge Data Eng.,* April 1995.

124. Dao, S. and Thuraisingham, B., Multimedia database systems, in *Advances in Database Systems,* Fortier, P., Ed., John Wiley and Sons, New York, NY, 1996.

125. Thuraisingham, B., Data warehousing, data mining, and security, Proceedings of the Tenth IFIP Database Security Conference, Como, Italy, July 1996.

126. Thuraisingham, B., *Data Management Systems Evolution and Interoperation,* CRC Press, Boca Raton, FL, 1997.

127. Thuraisingham, B., *Data Mining: Technologies, Techniques, Tools and Trends,* CRC Press, Boca Raton, FL, 1998.

128. Thuraisingham, B., *Web Information Management and Electronic Commerce,* CRC Press, Boca Raton, FL, 2000.

129. Special issue on multimedia database systems, *IEEE Trans. Knowledge Data Eng.,* April 2000.

130. Trueblood, R. and Potter, W., Hypersemantic data modeling, *Data Knowledge Eng. J.,* March 1988.

131. Tsur, D. et al., Query flocks: a generalization of association rule mining, Proceedings of the 1998 ACM SIGMOD Conference, Seattle, WA, June 1998.

132. Turban, E. and Aronson, J., *Decision Support Systems and Intelligent Systems,* Prentice-Hall, Englewood Cliffs, NJ, 1997.

133. Vazirgiannis, M. and Sellis, T., Multimedia database management systems, in *Advanced Database Technology and Design,* Piattini, M. and Diaz, O., Eds., Artech House, Norwood, MA, 2000.

134. *http://www.virage.com/products/vas.html.*

135. Grinstein, G., Ed., Proceedings of the 1995 Workshop on Visualization and Databases, Atlanta, GA, October 1997.

136. Grinstein, G., Ed., Proceedings of the 1995 Workshop on Visualization and Databases, Phoenix, AZ, October 1997.

137. Proceedings of the VLDB Conference, New York, N.Y., 1998.
138. Web Data Mining Workshop at the Knowledge Discovery in Databases Conference, San Diego, CA, August 1999.
139. Woelk, D. et al., An object-oriented approach to multimedia databases, Proceedings of the ACM SIGMOD Conference, Washington, D.C., June 1986.
140. *http://www.W3c.org*.
141. *http://www.xml.org/*.
142. Wolf, V., et al., Real-Time CORBA, *IEEE Transactions on Parallel and Distributed Systems,* October, 2000.
143. Ferrari, E. and Thuraisingham, B., Database security: survey, in *Advanced Database Technology and Design,* Piattini, M. and Diaz, O., Eds., Artech House, Norwood, MA, 2000.
144. Worldwide Web Consortium Report on Resource Description Format, Cambridge, MA, June 2000.
145. Fikes, R., Ontologies Presentation, Stanford University, December 1999.

Appendices

Appendix A
Data Management Systems: Developments and Trends

A.1 OVERVIEW

This appendix provides an overview of the developments and trends in data management as discussed by Thuraisingham.[9] Since data play a major role in Web data management, a good understanding of data management is essential for Web data management.

Recent developments in information systems technologies have resulted in the computerization of many applications in various business areas. Data have become a critical resource in many organizations, and, therefore, efficient access to data, sharing the data, extracting information from the data, and making use of the information have become urgent needs. As a result, there have been several efforts to integrate the various data sources scattered across several sites. These data sources may be databases managed by database management systems, or they could simply be files. To provide the interoperability between the multiple data sources and systems, various tools are being developed. These tools enable users of one system to access other systems in an efficient and easy manner.

We define data management systems to be systems that manage the data, extract meaningful information from the data, and make use of the information extracted. Therefore, data management systems include database systems, data warehouses, and data mining systems. Data could be structured, such as that found in relational databases, or it could be unstructured, such as text, voice, imagery, and video. There have been numerous discussions in the past to distinguish between data, information, and knowledge.* We do not attempt to clarify these terms. For our purposes, data could be just bits and bytes, or it could convey some meaningful information to the user. We will, however, distinguish between database systems and database management systems. A database management system is the component that manages the database containing persistent data. A database system consists of both the database and the database management system.

A key component to the evolution and interoperation of data management systems is the interoperability of heterogeneous database systems. Efforts regarding the interoperability between database systems have been reported since the late 1970s. However,

---

* Recently, the area of knowledge management has received a lot of attention. We addressed knowledge management in Chapter 13. More details are given by Morey.[5]

233

only recently are we seeing commercial developments in heterogeneous database systems. Major database system vendors are now providing interoperability between their products and other systems. Furthermore, many database system vendors are migrating towards an architecture called the client–server architecture, which facilitates distributed data management capabilities. In addition to efforts regarding the interoperability between different database systems and client–server environments, work is also being directed towards handling autonomous and federated environments.

The organization of this appendix is as follows. Since database systems are a key component of data management systems, we first provide an overview of the developments in database systems. These developments are discussed in Section A.2. We then provide a vision for data management systems in Section A.3. Our framework for data management systems is discussed in Section A.4. Note that data mining, warehousing, and Web data management are components of this framework. Building information systems from our framework with special instantiations is discussed in Section A.5. The relationship between the various texts written by this author is discussed in Section A.6. This appendix is summarized in Section A.7.

A.2   DEVELOPMENTS IN DATABASE SYSTEMS

Figure A.1 provides an overview of the developments in database systems technology. While early work in the 1960s focused on developing products based on network and hierarchical data models, many of the developments in database systems took place after the seminal paper by Codd describing the relational model[3] (see also Date[4]). Research and development work on relational database systems was carried out during the early 1970s, and several prototypes were developed throughout the 1970s. Notable efforts include IBM's System R and the University of California at Berkeley's Ingres. During the 1980s, many relational database system products were being marketed (notable among these products are those of Oracle Corporation, Sybase Inc., Informix Corporation, Ingres Corporation, IBM, Digital Equipment Corporation, and Hewlett Packard Company). During the 1990s, products from other vendors emerged (e.g., Microsoft Corporation). In fact, to date, numerous relational database system products have been marketed. However, Codd has stated that many of the systems marketed as relational systems are not really relational (see, for example, the discussion in Date[4]). He then discusses various criteria that a system must satisfy to be qualified as a relational database system. While early work focused on issues such as data models, normalization theory, query processing and optimization strategies, query languages, and access strategies and indexes, the focus later shifted toward supporting a multi-user environment. In particular, concurrency control and recovery techniques were developed. Support for transaction processing was also provided.

Research on relational database systems as well as transaction management was followed by research on distributed database systems around the mid-1970s. Several distributed database system prototype development efforts also began around the late 1970s. Notable among these efforts include IBM's System R, DDTS (distributed database testbed system) by Honeywell Inc., SDD-I and Multibase by CCA (Computer Corporation of America), and Mermaid by SDC (System Development Corporation). Furthermore, many of these systems (e.g., DDTS, Multibase, and Mermaid) function

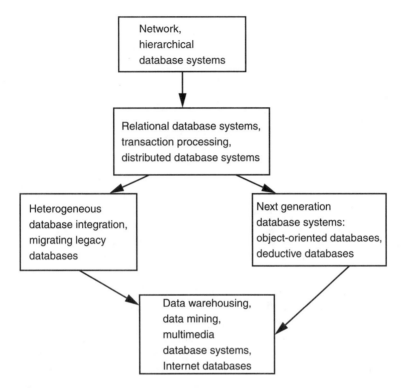

FIGURE A.1 Developments in database systems technology.

in a heterogeneous environment. During the early 1990s, several database system vendors, such as Oracle Corporation, Sybase Inc., and Informix Corporation, provided data distribution capabilities for their systems. Most of the distributed relational database system products are based on client–server architectures. The idea is to have the client of vendor A communicate with the server database system of vendor B. In other words, the client–server computing paradigm facilitates a heterogeneous computing environment. Interoperability between relational and nonrelational commercial database systems is also possible. The database systems community is also involved in standardization efforts. Notable among the standardization efforts are the ANSI/SPARC three-level schema architecture,* the IRDS (information resource dictionary system) standard for Data Dictionary Systems, the relational query language SQL (structured query language), and the RDA (remote database access) protocol for remote database access.

Another significant development in database technology is the advent of object-oriented database management systems. Active work on developing such systems began in the mid-1980s, and they are now commercially available (notable among them include the products of Object Design Inc., Ontos Inc., Gemstone Systems Inc., and Versant Object Technology). It was felt that new generation applications

---

* ANSI stands for American National Standards Institute. SPARC stands for Systems Planning and Requirements Committee.

such as multimedia, office information systems, CAD/CAM,* process control, and software engineering have different requirements. Such applications utilize complex data structures. Tighter integration between the programming language and the data model is also desired. Object-oriented database systems satisfy most of the requirements of these new generation applications.[2]

According to the Lagunita report published as a result of a National Science Foundation (NSF) workshop in 1990,[6,8] relational database systems, transaction processing, and distributed (relational) database systems are mature technologies. Furthermore, vendors are marketing object-oriented database systems and demonstrating the interoperability between different database systems. That report goes on to state that as applications are getting increasingly complex, more sophisticated database systems are needed. Furthermore, since many organizations now use database systems, in many cases of different types, the database systems need to be integrated. Although recent work has begun to address these issues and commercial products are available, several issues still need to be resolved. Therefore, challenges faced by the database systems researchers in the early 1990s were in two areas. One was next generation database systems, and the other was heterogeneous database systems.

Next generation database systems include object-oriented, functional, high performance, real-time, scientific, temporal, and intelligent database systems (also sometimes called logic or deductive database systems), special parallel architectures to enhance the performance of database system functions, and database systems that handle incomplete and uncertain information.** Ideally, a database system should provide the support for high performance transaction processing, model complex applications, represent new kinds of data, and make intelligent deductions. While significant progress was made during the late 1980s and early 1990s, there is still much to be done before such a database system can be developed.

Heterogeneous database systems received considerable attention during the last decade.[1] The major issues include handling different data models, different query processing strategies, different transaction processing algorithms, and different query languages. Should a uniform view be provided to the entire system, or should the users of the individual systems maintain their own views of the entire system? This is a question that has yet to be answered satisfactorily. A complete solution to heterogeneous database management systems is likely a generation away. While research should be directed towards finding such a solution, work should also be carried out to handle limited forms of heterogeneity to satisfy customer needs. Another type of database system that has received some attention lately is a federated database system. Note that some have used the terms heterogeneous and federated interchangeably. While heterogeneous database systems can be part of a federation, a federation can also include homogeneous database systems.

The explosion of users on the Internet and the Web as well as developments in interface technologies have resulted in even more challenges for data management researchers. A second workshop was sponsored by the NSF in 1995, and several

---

* CAD/CAM stands for Computer Aided Design/Computer Aided Manufacturing.
** For a discussion of the next generation database systems, refer to the *ACM SIGMOD Record,* December 1980.[8]

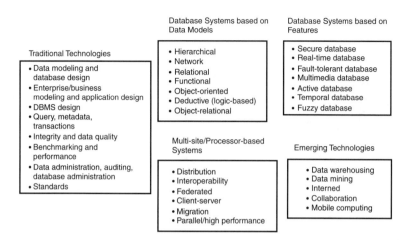

FIGURE A.2 Comprehensive view of data management systems.

emerging technologies have been identified as important in the twenty-first century.[7] These include digital libraries, managing very large databases, data administration issues, multimedia databases, data warehousing, data mining, data management for collaborative computing environments, and security and privacy. Another significant development of the 1990s was the development of object-relational systems. Such systems combine the advantages of both object-oriented and relational database systems. Also, many corporations are now focusing on integrating their data management products with Internet technologies. Finally, for many organizations, there is an increasing need to migrate some of the legacy databases and applications to newer architectures and systems such as client–server architectures and relational database systems. We believe there is no end to data management systems. As new technologies are developed, there are new opportunities for data management research and development.

A comprehensive view of all data management technologies is illustrated in Figure A.2. As shown, traditional technologies include database design, transaction processing, and benchmarking. Then there are database systems based on data models such as relational and object-oriented. Database systems may depend on features they provide, such as security and real-time. These database systems may be relational or object-oriented. There are also database systems based on multiple sites or processors, such as distributed and heterogeneous database systems, parallel systems, and systems being migrated. Finally, there are the emerging technologies such as data warehousing and mining, collaboration, and the Internet. Any comprehensive text on data management systems should address all of these technologies. We have selected some of the relevant technologies and put them in a framework, which is described in Section A.5.

A.3 STATUS, VISION, AND ISSUES

Significant progress has been made on data management systems. However, many of the technologies are still stand-alone, as illustrated in Figure A.3. For example, multimedia systems are yet to be successfully integrated with warehousing and mining technologies. The ultimate goal is to integrate multiple technologies so that

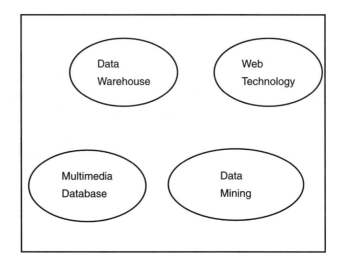

FIGURE A.3 Stand-alone systems.

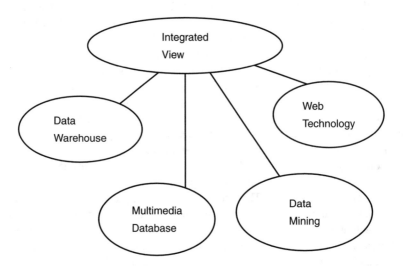

FIGURE A.4 Vision.

accurate data and information are produced at the right time and distributed to the user in a timely manner. Our vision for data and information management is illustrated in Figure A.4.

The work discussed by Thuraisingham[9] addressed many of the challenges necessary to accomplish this vision. In particular, integration of heterogeneous databases as well as the use of distributed object technology for interoperability were discussed. While much progress has been made on the system aspects of interoperability, semantic issues still remain a challenge. Different databases have different representations. Furthermore, the same data entity may be interpreted differently at different sites.

Addressing these semantic differences and extracting useful information from the heterogeneous and possibly multimedia data sources are major challenges. This book has attempted to address some of the challenges through the use of data mining.

A.4 DATA MANAGEMENT SYSTEMS FRAMEWORK

For the successful development of evolvable interoperable data management systems, heterogeneous database systems integration is a major component. However, there are other technologies that have to be successfully integrated with each other to develop techniques for efficient access and sharing of data as well as for the extraction of information from the data. To facilitate the development of data management systems to meet the requirements of various applications in fields such as medicine, finance, manufacturing, and military, we have proposed a framework, which can be regarded as a reference model, for data management systems. Various components of this framework have to be integrated to develop data management systems to support various applications.

Figure A.5 illustrates our framework, which can be regarded as a model, for data management systems.* This framework consists of three layers. One can think of the component technologies, which are also referred to as components, belonging to a particular layer as more or less built upon the technologies provided by the lower layer. Layer I is the database technology and distribution layer, which consists of database systems and distributed database systems technologies. Layer II is the interoperability and migration layer, which consists of technologies such as heterogeneous database integration, client–server databases, multimedia database systems to handle heterogeneous data types, and migrating legacy databases.** Layer III is the information extraction and sharing layer, which essentially consists of technologies for some of the newer services supported by data management systems. These include data warehousing, data mining,[10] Internet databases, and database support for collaborative applications.***,**** Data management systems may utilize lower level technologies such as networking, distributed processing, and mass storage. We have grouped these technologies into a layer called the supporting technologies layer, which does not belong to the data management systems framework. This supporting layer also consists of some higher-level technologies such as distributed object

---

* Note that this three-layer model is subjective and not a standard model. This model has helped us organize our views on data management.

** We have placed multimedia database systems in Layer II, as we consider them a special type of a heterogeneous database system. A multimedia database system handles heterogeneous data types such as text, audio, and video. That is the subject of this book.

*** Note that one can argue whether database support for collaborative applications should be discussed here because collaborative computing is not part of the data management framework. However, such applications do need database support, and our focus has been on this support.

**** Although Internet database management is an integration of various technologies, we have placed it in Layer III because it still deals with information extraction. Note that the data management framework consists of technologies for managing data as well as for extracting information from the data. However, what one does with the information, such as collaborative computing, sophisticated human computer interaction, natural language processing, and knowledge-based processing, does not belong to this framework. Those belong to the application technologies layer.

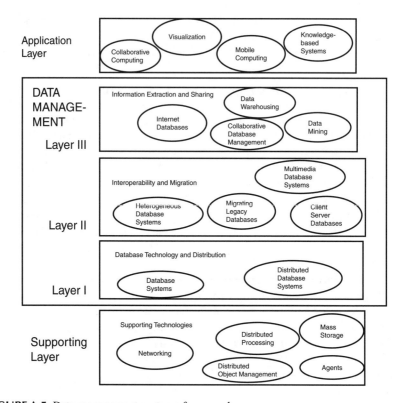

FIGURE A.5 Data management systems framework.

management and agents.* Figure A.5 also shows the application technologies layer. Systems such as collaborative computing systems and knowledge-based systems that belong to the application technologies layer may utilize data management systems. Note that the application technologies layer is also outside of the data management systems framework.

The technologies that constitute the data management systems framework can be regarded as some of the core technologies in data management. However, features like security, integrity, real-time processing, fault tolerance, and high performance computing are needed for many applications utilizing data management technologies. Applications utilizing data management technologies may be medical, financial, or military, among others. This is illustrated in Figure A.6, where a three-dimensional view relating data management technologies with features and applications is given. For example, one could develop a secure distributed database management system for medical applications or a fault tolerant multimedia database management system for financial applications.**

---

* Note that technologies such as distributed object management enable interoperation and migration.
** In some cases, one could also consider multimedia data processing and reengineering, which is an essential part of system migration, to be at the same level as features like security and integrity. One can also regard them as emerging technologies.

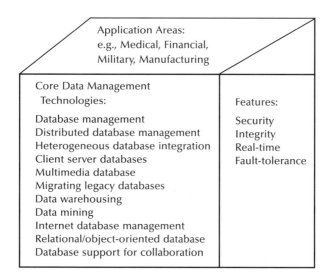

FIGURE A.6 A three-dimensional view of data management.

Integrating the components belonging to the various layers is important for developing efficient data management systems. In addition, data management technologies have to be integrated with application technologies to develop successful information systems. However, at present, there is limited integration between these various components. Our previous book focused mainly on the concepts, developments, and trends belonging to each of the components shown in the framework.[9] Furthermore, our current book on Web data management, which we also refer to as Internet data management, focuses on the Internet database component of Layer III of Figure A.5.

A.5 BUILDING INFORMATION SYSTEMS FROM THE FRAMEWORK

Figure A.5 illustrates a framework for data management systems. As shown in that figure, the technologies for data management include database systems, distributed database systems, heterogeneous database systems, migrating legacy databases, multimedia database systems, data warehousing, data mining, Internet databases, and database support for collaboration. Furthermore, data management systems take advantage of supporting technologies such as distributed processing and agents. Similarly, application technologies such as collaborative computing, visualization, expert systems, and mobile computing take advantage of data management systems.*

Many of us have heard the term information systems on numerous occasions. These systems have sometimes been used interchangeably with data management systems. In our terminology, information systems are much broader than and include data management systems. In fact, a framework for information systems will include

---

* Note that databases can also support expert systems, as in the case of collaborative applications.

Collaboration, Visualization

Multimedia Database, Distributed Database Systems

Mass Storage, Distributed Processing

FIGURE A.7 Framework for multimedia data management for collaboration.

not only the data management system layers, but also the supporting technologies and application technologies layers. That is, information systems encompass all kinds of computing systems. They can be regarded as the finished product that can be used for various applications. That is, while hardware is at the lowest end of the spectrum, applications are at the highest end.

We can combine the technologies of Figure A.5 to put together information systems. For example, at the application technology level, one may need collaboration and visualization technologies so that analysts can collaboratively carry out some tasks. At the data management level, one may need both multimedia and distributed database technologies. At the supporting level, one may need mass storage as well as some distributed processing capability. This special framework is illustrated in Figure A.7. Another example is a special framework for interoperability. One may need some visualization technology to display the integrated information from heterogeneous databases. At the data management level, we have heterogeneous database systems technology. At the supporting technology level, one may use distributed object management technology to encapsulate the heterogeneous databases. This special framework is illustrated in Figure A.8.

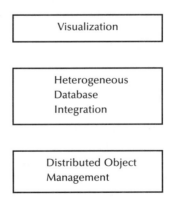

FIGURE A.8 Framework for heterogeneous database interoperability.

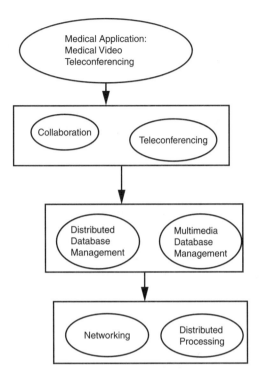

FIGURE A.9 Specific example.

Finally, let us illustrate the concepts that we have described above by using a specific example. Suppose a group of physicians/surgeons want a system where they can collaborate and make decisions about various patients. This system could be a medical video teleconferencing application. That is, at the highest level, the application is a medical application and, more specifically, a medical video teleconferencing application. At the application technology level, one needs a variety of technologies, including collaboration and teleconferencing. These application technologies will make use of data management technologies such as distributed and multimedia database systems. One may need to support multimedia data such as audio and video. The data management technologies, in turn, draw upon lower level technologies such as distributed processing and networking, as illustrated in Figure A.9.

In summary, information systems include data management systems as well as application-layer systems, such as collaborative computing systems, and supporting-layer systems, such as distributed object management systems.

While application technologies make use of data management technologies and data management technologies make use of supporting technologies, the ultimate user of the information system is the application itself. Today, numerous applications make use of information systems. These applications are from multiple domains such as medical, financial, manufacturing, telecommunications, and defense. Specific applications include signal processing, electronic commerce, patient monitoring, and

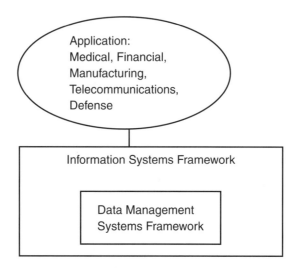

FIGURE A.10 Application–framework relationship.

situation assessment. Figure A.10 illustrates the relationship between the application and the information system.

A.6 RELATIONSHIPS BETWEEN THE TEXTS

We have published four books on data management and mining and are currently writing one more. These books are *Data Management Systems Evolution and Inter-operation*,[9] *Data Mining Technologies, Techniques, Tools and Trends*,[10] *Web Data Management and Electronic Commerce*,[11] *Managing and Mining Multimedia Data-bases for the Electronic Enterprise* (this book), and *XML, The Semantic Web, and E-Business*,[12] which is expected to be published later in 2001. All of these books have evolved from the framework that we illustrated in this appendix and address different parts of the framework. The connection between these texts is illustrated in Figure A.11.

A.7 SUMMARY

This appendix has provided an overview of data management. We first discussed the developments in data management and then provided a vision for data management. Then we illustrated a framework for data management. This framework consists of three layers: the database systems layer, interoperability layer, and information extraction layer. Web data management belongs to Layer III. Finally, we showed how information systems could be built from the technologies of the framework.

Let us repeat what we mentioned in Chapter 1 now that we have described the data management framework introduced by Thuraisingham.[9] This book not only

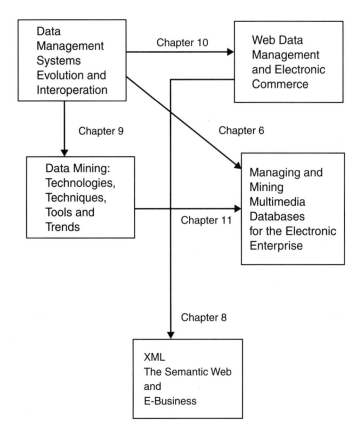

FIGURE A.11 Relationship between texts.

discussed multimedia data management and mining concepts, it also showed how multimedia data management and mining can be applied to the various applications such as electronic commerce. Many of the technologies discussed in the framework of Figure A.5 have been useful in the discussion of multimedia data management and mining. These include database systems, distributed database systems, data warehousing, and data mining. In addition, some other features for data management such as metadata and security also play a role in various chapters of this book. For example, metadata for multimedia databases was the subject of Chapter 4. Security and privacy issues were discussed in Chapter 10 with respect to the Web as well as with respect to multimedia data management and mining.

We believe that data management is essential to many information technologies including data mining, multimedia information processing, interoperability, and collaboration and knowledge management. This appendix stresses data management. The remaining appendices focus on various other key technologies for multimedia information processing including database systems, data mining, object technology, and the Web.

REFERENCES

1. Special Issue on Heterogeneous Database Systems, *ACM Comput. Surv.,* September 1990.
2. Cattell, R., *Object Data Management Systems,* Addison-Wesley, Reading, MA, 1991.
3. Codd, E. F., A relational model of data for large shared data banks, *Commun. ACM,* May 1970.
4. Date, C. J., *An Introduction to Database Management Systems,* Addison-Wesley, Reading, MA, 1990.
5. Morey, D., Knowledge management architecture, *Handbook of Data Management,* Thuraisingham, B., Ed., Auerbach Publications, New York, NY, 1998.
6. Proceedings of the Database Systems Workshop, Report published by the National Science Foundation, March 1990.
7. Proceedings of the Database Systems Workshop, Report published by the National Science Foundation, March 1995.
8. Next generation database systems, *ACM SIGMOD Rec.,* December 1990.
9. Thuraisingham, B., *Data Management Systems Evolution and Interoperation,* CRC Press, Boca Raton, FL, 1997.
10. Thuraisingham, B., *Data Mining: Technologies, Techniques, Tools and Trends,* CRC Press, Boca Raton, FL, 1998.
11. Thuraisingham, B., *Web Data Management and Electronic Commerce,* CRC Press, Boca Raton, FL, 2000.
12. Thuraisingham, B., *XML, The Semantic Web and E-Business,* CRC Press, Boca Raton, FL, in press.

Appendix B
Database Systems Technology

B.1 OVERVIEW

Database systems play a key role in Web data management. Having good data is key to effective Web data management, and, therefore, we give considerable attention to database systems in this book. It should be noted that we are taking a data-oriented approach to the Web.

Database systems technology has advanced a great deal during the past four decades from the legacy systems based on network and hierarchical models to relational and object-oriented database systems based on client–server architectures. This appendix provides an overview of the important developments in database systems relevant to the contents of this book. Much of the discussion in this book builds on the information presented in this appendix.

As stated in Appendix A, we consider a database system to include both the database management system (DBMS) and the database (see also the discussion of Date[14]). The DBMS component of the database system manages the database. The database contains persistent data. That is, the data are permanent even if the application programs go away.

The organization of this appendix is as follows. In Section B.2, relational data models as well as entity relationship models are discussed. In Section B.3, various types of architectures for database systems are described. These include an architecture for a centralized database system, schema architecture, as well as functional architecture. Database design issues are discussed in Section B.4. Database administration issues are discussed in Section B.5. Section B.6 addresses database system functions, including query processing, transaction management, metadata management, storage management, maintaining integrity and security, and fault tolerance. While the previous sections discuss database functions for a centralized database system, the remaining sections address distribution, interoperability and migration. For example, distributed database systems are the subject of B.7. Heterogeneous database integration aspects are summarized in section B.8. Managing federated databases is the subject of Section B.9. Migrating legacy databases is discussed in Section B.10. Client–server database management is the subject of Section B.11. The impact of the Web is the subject of Section B.12. The appendix is summarized in Section B.13.*

---

* There are other data management technologies. For example, data warehousing is discussed in the appendix on data mining. Distributed, heterogeneous, and client–server data management are discussed later in this appendix.

B.2 RELATIONAL AND ENTITY-RELATIONSHIP DATA MODELS

B.2.1 OVERVIEW

It is widely accepted among the data modeling community that the purpose of a data model is to capture the universe that it is representing as accurately, completely, and naturally as possible.[42] Various data models have been proposed, and we have provided an overview in a previous book.[39] This section discusses the essential points of the relational data model, as it is the most widely used today. In addition, we also discuss entity-relationship data models, as some of the ideas have been used in object models and, furthermore, entity-relationship models are being used extensively in database design. Many other models exist as well, such as logic-based, hypersemantic, and functional models. Discussion of all of these models is beyond the scope of this book. We do provide an overview of an object model in Appendix D, as object technology is essential for the Web.[28,32]

B.2.2 RELATIONAL DATA MODEL

In the relational model,[12] the database is viewed as a collection of relations. Each relation has attributes and rows. For example, Figure B.1 illustrates a database with two relations, EMP and DEPT. EMP has four attributes: SS#, Ename, Salary, and D#. DEPT has three attributes: D#, Dname, and Mgr. EMP has three rows, also called tuples, and DEPT has two rows. Each row is uniquely identified by its primary key. For example, SS# could be the primary key for EMP, and D# could be the primary key for DEPT. Another key feature of the relational model is that each element in the relation is an atomic value such as an integer or a string. That is, complex values such as lists are not supported.

Various operations are performed on relations. The select operation selects a subset of rows satisfying certain conditions. For example, in the relation EMP, one may select tuples where the salary is more than $20,000. The project operation projects the relation onto some attributes. For example, in the relation EMP, one may project onto the attributes Ename and Salary. The join operation joins two relations over some common attributes. A detailed discussion of these operations is given in Date[14] and Ullman.[43]

EMP

SS#	Ename	Salary	D#
1	John	20K	10
2	Paul	30K	20
3	Mary	40K	20

DEPT

D#	Dname	Mgr
10	Math	Smith
20	Physics	Jones

FIGURE B.1 Relational database.

FIGURE B.2 Entity-relationship representation.

Various languages to manipulate the relations have been proposed. Notable among these languages is the ANSI Standard SQL, which is used to access and manipulate data in relational databases.[38] There is wide acceptance of this standard among database management system vendors and users. It supports schema definition, retrieval, data manipulation, schema manipulation, transaction management, integrity, and security. Other languages include the relational calculus first proposed in the Ingres project at the University of California at Berkeley.[14] Another important concept in relational databases is the notion of a view. A view is essentially a virtual relation and is formed from the relations in the database. For further details, refer to Date.[14]

B.2.3 ENTITY-RELATIONSHIP (ER) DATA MODEL

One of the major drawbacks of the relational data model is its lack of support for capturing the semantics of an application. This resulted in the development of semantic data models. The entity-relationship (ER) data model developed by Chen[10] can be regarded as the earliest semantic data model. In this model, the world is viewed as a collection of entities and relationships between entities. Figure B.2 illustrates two entities, EMP and DEPT. The relationship between them is WORKS.

Relationships can be either one–one, many–one, or many–many. If each employee works in one department and each department has one employee, then WORKS is a one–one relationship. If an employee works in one department and each department can have many employees, then WORKS is a many–one relationship. If an employee works in many departments, and each department has many employees, then WORKS is a many–many relationship.

Several extensions to the entity-relationship model have been proposed. One is the entity-relationship-attribute model, where attributes are associated with entities as well as relationships, and another has introduced the notion of categories into the model.[21,45] ER models are used mainly to design databases. That is, most database computer aided software engineering (CASE) tools are based on the ER model, where the application is represented using such a model, and, subsequently, the database (possibly relational) is generated. Current database management systems are not based on the ER model. That is, unlike the relational model, ER models did not take off in the development of database management systems.

B.3 ARCHITECTURAL ISSUES

This section describes various types of architectures for a database system. First, we illustrate a very high-level centralized architecture for a database system. We then describe a functional architecture for a database system. In particular, the

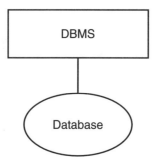

FIGURE B.3 Centralized architecture.

FIGURE B.4 Functional architecture for a DBMS.

functions of the DBMS component of the database system are illustrated in this architecture. Then we discuss the ANSI/SPARC's three-schema architecture, which has been more or less accepted by the database community.[14] Finally, we describe extensible architectures.*

Figure B.3 is an example of a centralized architecture. In that figure, the DBMS is a monolithic entity and manages a database which is centralized. Functional architecture illustrates the functional modules of a DBMS. The major modules of a DBMS include the query processor, transaction manager, metadata manager, storage manager, integrity manager, and security manager. The functional architecture of the DBMS component of the centralized database system architecture (of Figure B.3) is illustrated in Figure B.4.

Schema describes the data in the database. It has also been referred to as the data dictionary or contents of the metadatabase. Three-schema architecture was proposed for a centralized database system in the 1960s. This is illustrated in Figure B.5. The levels are the external schema, which provides an external view, the conceptual schema, which provides a conceptual view, and the internal schema, which provides an internal view. Mappings between the different schemas must be provided to transform one representation into another. For example, at the external

---

* Note that distributed architectures for data management are discussed in Appendix D where we address distributed, heterogeneous, and legacy databases.

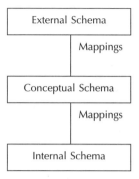

FIGURE B.5 Three-schema architecture.

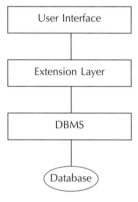

FIGURE B.6 Extensible DBMS.

level, one could use ER representation. At the logical or conceptual level, one could use relational representation. At the physical level, one could use a representation based on B trees.*

There is also another aspect to architectures, and that is extensible database architectures. For example, for many applications, a DBMS may have to be extended with a layer to support objects, process rules, handle multimedia data types, or even do mining. Such an extensible architecture is illustrated in Figure B.6.

B.4 DATABASE DESIGN

Designing a database is a complex process. Much of the work that has been done has been on designing relational databases. There are three steps involved, which are illustrated in Figure B.7. The first step is to capture the entities of the application and the relationships between the entities. One could use a model such as the entity-relationship model for this purpose. More recently, object-oriented data models,

---

* A B tree is a representation scheme used to physically represent the data. However, it is at a higher level than the bits and bytes level. For a discussion of physical structures and models, refer to Date.[14]

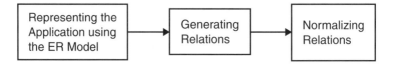

FIGURE B.7 Database design process.

which are part of object-oriented design and analysis methodologies, are becoming popular to represent the application.

The second step is to generate the relations from the representations. For example, from the entity-relationship diagram of Figure B.2, one could generate the relations EMP, DEPT, and WORKS. The relation WORKS will capture the relationship between employees and departments.

The third step is to design good relations. This is the normalization process. Various normal forms have been defined in the literature (see, for example, Maier[29] and Date[14]). For many applications, relations in the third normal form would suffice. With this normal form, redundancies, complex values, and other situations that could cause potential anomalies are eliminated.

B.5 DATABASE ADMINISTRATION

Every database has a database administrator (DBA). It is the responsibility of the DBA to define the various schemas and mappings. In addition, the functions of the administrator include auditing the database as well as implementing appropriate backup and recovery procedures.

The DBA could also be responsible for maintaining the security of the system. In some cases, the system security officer (SSO) maintains security. The administrator should determine the granularity of the data for auditing. For example, in some cases, there is tuple (or row) level auditing, while in other cases there is table (or relation) level auditing. It is also the administrator's responsibility to analyze the audit data.

Note that there is a difference between database administration and data administration. Database administration assumes there is an installed database system. The DBA manages this system. Data administration functions include conducting data analysis, determining how a corporation handles its data, and enforcing appropriate policies and procedures for managing the data of a corporation. Data administration functions are carried out by the data administrator. For a discussion of data administration, refer to von Halle and Kull,[15,16] Thuraisingham,[17,18] and Proceedings of the DoD Database Colloquium.[19,20] Figure B.8 illustrates various database administration issues.

B.6 DATABASE MANAGEMENT SYSTEM FUNCTIONS

B.6.1 OVERVIEW

The functional architecture of a DBMS was illustrated in Figure B.4. The functions of a DBMS carry out its operations. A DBMS essentially manages a database and

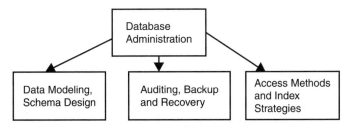

FIGURE B.8 Some database administration issues.

provides support to the user by enabling him to query and update the database. Therefore, the basic functions of a DBMS are query processing and update processing. In some applications, such as banking, queries and updates are issued as part of transactions. Therefore, transaction management is also another function of a DBMS. To carry out these functions, information about the database has to be maintained. This information is called metadata. The function that is associated with managing the metadata is metadata management. Special techniques are needed to manage the data stores that actually store the data. The function that is associated with managing these techniques is storage management. To ensure that the above functions are carried out properly and that the user gets accurate data, there are some additional functions. These include security management, integrity management, and fault management (i.e., fault tolerance).

The above are some of the essential functions of a DBMS. However, there has recently been emphasis on extracting information from the data. Therefore, other functions of a DBMS may include providing support for data mining, data warehousing, and collaboration.

This section focuses only on the essential functions of a DBMS. These are query processing, transaction management, metadata management, storage management, maintaining integrity, security control, and fault tolerance. Note that we do not have a special section for update processing, as it can be done as part of transaction management. We discuss each of the essential functions in Sections B.6.2 through B.6.7.

B.6.2 QUERY PROCESSING

Query operation is the most commonly used function in a DBMS. It should be possible for users to query the database and obtain answers to their queries. There are several aspects to query processing. First of all, a good query language is needed. Languages such as SQL are popular for relational databases. Such languages are being extended for other types of databases. The second aspect is techniques for query processing. Numerous algorithms have been proposed for query processing in general and for the joint operation in particular.[25] Also, different strategies are possible to execute a particular query. The costs for the various strategies are computed, and the one with the least cost is usually selected for processing. This process is called query optimization. Cost is generally determined by the disk access. The goal is to minimize disk access in processing a query.

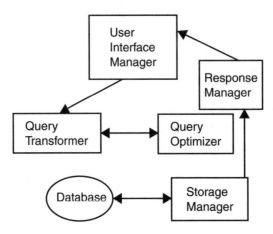

FIGURE B.9 Query processor.

As stated earlier, users pose a query using a language. The constructs of the language have to be transformed into the constructs understood by the database system. This process is called query transformation. Query transformation is carried out in stages based on the various schemas. For example, a query based on the external schema is transformed into a query on the conceptual schema. This is then transformed into a query on the physical schema. In general, rules used in the transformation process include the factoring of common subexpressions and pushing selections and projections down in the query tree as much as possible. If selections and projections are performed before the joins, then the cost of the joins can be considerably reduced.

Figure B.9 illustrates the modules in query processing. The user interface manager accepts queries, parses the queries, and then gives them to the query transformer. The query transformer and query optimizer communicate with each other to produce an execution strategy. The database is accessed through the storage manager. The response manager gives responses to the user.

B.6.3 TRANSACTION MANAGEMENT

A transaction is a program unit that must be executed in its entirety or not executed at all. If transactions are executed serially, then there is a performance bottleneck. Therefore, transactions are executed concurrently. Appropriate techniques must ensure that the database is consistent when multiple transactions update the database. That is, transactions must satisfy the ACID (atomicity, consistency, isolation, and durability) properties. Major aspects of transaction management are serializability, concurrency control, and recovery. We discuss them briefly in this section. For a detailed discussion of transaction management, refer to Date[14] and Ullman.[43] A good theoretical treatment of this topic is given by Bernstein et al.[4]

A schedule is a sequence of operations performed by multiple transactions. Two schedules are equivalent if their outcomes are the same. A serial schedule is a schedule where no two transactions execute concurrently. One objective of transaction

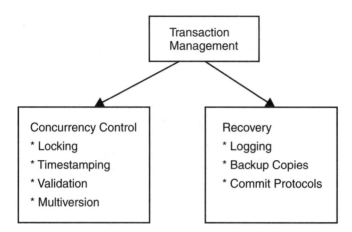

FIGURE B.10 Some aspects of transaction management.

management is to ensure that any schedule is equivalent to a serial schedule. Such a schedule is called a serializable schedule. Various conditions for testing the serializability of a schedule have been formulated for a DBMS.

Concurrency control techniques ensure that the database is in a consistent state when multiple transactions update the database. Three popular concurrency control techniques which ensure the serializability of schedules are locking, time-stamping, and validation.

If a transaction aborts due to some failure, then the database must be brought to a consistent state, which is called transaction recovery. One solution to handling transaction failure is to maintain log files. The transaction's actions are recorded in the log file. So, if a transaction aborts, the database is brought back to a consistent state by undoing the actions of the transaction. The information for the undo operation is found in the log file. Another solution is to record the actions of a transaction but not make any changes to the database. Only if a transaction commits should the database be updated. There are some issues, however. For example, the log files have to be kept in stable storage. Various modifications to the above techniques have been proposed to handle the various situations.

When transactions are executed at multiple data sources, then a protocol called two-phase commit is used to ensure that the multiple data sources are consistent. Figure B.10 illustrates the various aspects of transaction management.

B.6.4 Storage Management

The storage manager is responsible for accessing the database. To improve the efficiency of query and update algorithms, appropriate access methods and index strategies have to be enforced. That is, in generating strategies for executing query and update requests, the access methods and index strategies that are used need to be taken into consideration. The access methods used to access the database would depend on the indexing methods. Therefore, creating and maintaining appropriate index files is a major issue in database management systems. By using an appropriate

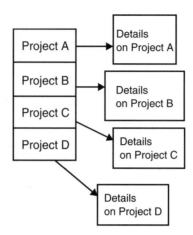

FIGURE B.11 An example index on projects.

indexing mechanism, the query processing algorithms may not have to search the entire database. Instead, the data to be retrieved could be accessed directly. Consequently, the retrieval algorithms are more efficient. Figure B.11 illustrates an example of an indexing strategy where the database is indexed by projects.

Much research has been carried out on developing appropriate access methods and index strategies for relational database systems. Some examples of index strategies are B trees and hashing.[14] Current research focuses on developing such mechanisms for object-oriented database systems with support for multimedia data.

B.6.5 METADATA MANAGEMENT

Metadata describes the data in a database. For example, in the case of the relational database illustrated in Figure B.1, metadata would include the following information: the database has two relations, EMP and DEPT; EMP has four attributes and DEPT has three attributes, etc. One of the main issues is developing a data model for metadata. In our example, one could use a relational model to model the metadata too. The metadata relation REL shown in Figure B.12 consists of information about relations and attributes.

In addition to information about the data in the database, metadata also includes information on access methods, index strategies, security constraints, and integrity constraints. One could also include policies and procedures as part of the metadata. In other words, there is no standard definition for metadata. There are, however, efforts to standardize metadata.[30] Metadata becomes a major issue with some of the recent developments in data management, e.g., digital libraries. Some of the issues are discussed in Part II of this book.

Once the metadata is defined, the issues include managing the metadata. What are the techniques for querying and updating the metadata? Since all of the other DBMS components need to access the metadata for processing, what are the interfaces between the metadata manager and the other components? Metadata management is fairly well understood for relational database systems. The current challenge

Relation REL

Relation	Attribute
EMP	SS#
EMP	Ename
EMP	Salary
EMP	D#
DEPT	D#
DEPT	Dname
DEPT	Mgr

FIGURE B.12 Metadata relation.

is to manage the metadata for more complex systems such as digital libraries and Internet database systems.

B.6.6 DATABASE INTEGRITY

Concurrency control and recovery techniques maintain the integrity of the database. In addition, there is another type of database integrity, i.e., enforcing integrity constraints. There are two types of integrity constraints enforced in database systems: application-independent integrity constraints and application-specific integrity constraints. Integrity mechanisms also include techniques for determining the quality of the data. For example, what is the accuracy of the data and the source? What are the mechanisms for maintaining the quality of the data? How accurate is the data on output? For a discussion of integrity based on data quality, refer to the MIT Technical Reports on data quality.[31] Note that data quality is very important for mining and warehousing. If the data that is mined is not good, one cannot rely on the results. That idea is discussed further in Appendix E.

Application-independent integrity constraints include the primary key constraint, the entity integrity rule, the referential integrity constraint, and the various functional dependencies involved in the normalization process.[14]

Application-specific integrity constraints are those constraints that are specific to an application. Examples include, "an employee's salary cannot decrease," and "no manager can manage more than two departments." Various techniques have been proposed to enforce application specific integrity constraints. For example, when the database is updated, these constraints are checked and the data are validated. Aspects of database integrity are illustrated in Figure B.13.

B.6.7 DATABASE SECURITY

This section focuses on discretionary security because this is an area of interest with respect to warehousing and mining. The major issues in security are authentication, identification, and enforcing appropriate access controls. For example, what are the mechanisms for identifying and authenticating a user? Will simple password mechanisms suffice? With respect to access control rules, languages such as SQL have

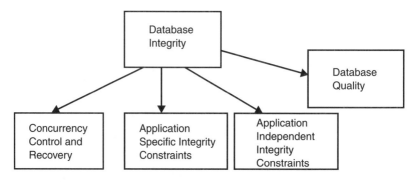

FIGURE B.13 Some aspects of database integrity.

Members of Group A can access all
information in employee database
except salary values

John, who is a member of Group A,
has access to salary values

Name and salary values taken
together can only be accessed by
members of Group B

FIGURE B.14 Access control rules.

incorporated grant and revoke statements to grant and revoke access to users. For many applications, simple grant and revoke statements are not sufficient. There may be more complex authorizations based on database content. Negative authorizations may also be needed. Access to data based on the roles of the user is also being investigated.

Numerous papers have been published on discretionary security in databases. These can be found in various security-related journals and conference proceedings.[24] Some aspects of database security are illustrated in Figure B.14.

B.6.8 Fault Tolerance

The previous two sections discussed database integrity and security. A closely related feature is fault tolerance. It is almost impossible to guarantee that the database will function as planned. In reality, various faults may occur. These could be hardware or software faults. As mentioned earlier, one of the major issues in transaction management is to ensure that the database is brought back to a consistent state in the presence of faults. The solutions proposed include maintaining appropriate log files to record the actions of a transaction in case its actions have to be retraced.

Another approach to handling faults is checkpointing. Various checkpoints are placed during the course of database processing. At each checkpoint, it is ensured that the database is in a consistent state. Therefore, if a fault occurs during processing,

```
Checkpoint A
Start Processing
*

*

Acceptance Test
If OK, then go to Checkpoint B
Else Roll Back to Checkpoint A

Checkpoint B
Start Processing
*

*
```

FIGURE B.15 Some aspects of fault tolerance.

then the database must be brought back to the last checkpoint. That way, the database can be guaranteed to be consistent. Closely associated with checkpointing are acceptance tests. After various processing steps, the acceptance tests are checked. If the techniques pass the tests, they can proceed further. Some aspects of fault tolerance are illustrated in Figure B.15.

B.7 DISTRIBUTED DATABASES

Although many definitions of a distributed database system have been given, no definition is standard. Our discussion of distributed database system concepts and issues has been influenced by the discussion in Ceri and Pelagatti.[9] A distributed database system includes a distributed database management system (DDBMS), a distributed database, and a network for interconnection. The DDBMS manages the distributed database. A distributed database is data that are distributed across multiple databases. Our choice architecture for a distributed database system is a multi-database architecture, which is tightly coupled. This architecture is illustrated in Figure B.16. We have chosen such an architecture because we can explain the concepts for both homogeneous and heterogeneous systems based on this approach. In this architecture, the nodes are connected via a communication subsystem, and local applications are handled by the local DBMS. In addition, each node is also involved in at least one global application, so there is no centralized control in this architecture. The DBMSs are connected through a component called the distributed processor (DP). In a homogeneous environment, the local DBMSs are homogeneous, while in a heterogeneous environment, the local DBMSs may be heterogeneous.

Distributed database system functions include distributed query processing, distributed transaction management, distributed metadata management, and enforcing security and integrity across the multiple nodes.[2] The DP is a critical component of the DDBMS; it connects the different local DBMSs. That is, each local DBMS is

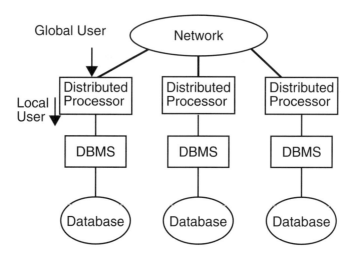

FIGURE B.16 An architecture for a DDBMS.

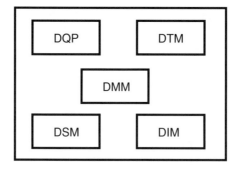

FIGURE B.17 Modules of a DP.

augmented by a DP. The modules of the DP are illustrated in Figure B.17. The components are the distributed metadata manager (DMM), the distributed query processor (DQP), the distributed transaction manager (DTM), the distributed security manager (DSP), and the distributed integrity manager (DIM). The DMM manages the global metadata. The global metadata includes information on the schemas which describe the relations in the distributed database, the way the relations are fragmented, the locations of the fragments, and the constraints enforced. The DQP is responsible for distributed query processing; the DTM is responsible for distributed transaction management; the DSM is responsible for enforcing global security constraints; and the DIM is responsible for maintaining integrity at the global level. Note that the modules of the DP communicate with their peers at the remote nodes. For example, the DQP at node 1 communicates with the DQP at node 2 for handling distributed queries.

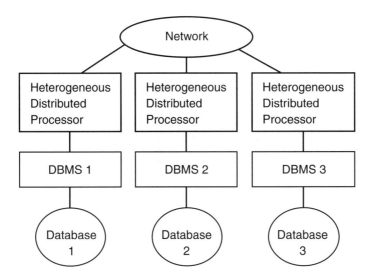

FIGURE B.18 Interoperability of heterogeneous database systems.

B.8 HETEROGENEOUS DATABASE INTEGRATION

Figure B.18 illustrates an example of interoperability between heterogeneous database systems. The goal is to provide transparent access, both for users and application programs, for querying and executing transactions.[1,23,44] Note that in a heterogeneous environment, the local DBMSs may be heterogeneous. Furthermore, the modules of the DP have both local DBMS-specific processing as well as local DBMS-independent processing. We call such a DP a heterogeneous distributed processor (HDP). Some of these issues are discussed in more detail by Thuraisingham.[39]

There are several technical issues that need to be resolved for the successful interoperation between these diverse database systems. Note that heterogeneity could exist with respect to different data models, schemas, query processing techniques, query languages, transaction management techniques, semantics, integrity, and security. There are two approaches to interoperability. One is the federated database management approach, in which cooperating, autonomous, and possibly heterogeneous component database systems, each belonging to one or more federations, communicate with each other. The other is the client–server approach, in which the goal is for multiple clients to communicate with multiple servers in a transparent manner.

Our previous book, *Data Management Systems Evolution and Interoperation*, addresses both aspects of interoperability.[39] Various aspects of heterogeneity are also addressed in that book. We are often asked when one should interconnect heterogeneous database systems through an HDP and when one should integrate them through a data warehouse. For on-line transaction processing applications, the interoperability approach is the answer, whereas for decision support applications, the warehousing approach is the answer. For other applications, one may need both, as illustrated in Figure B.19.

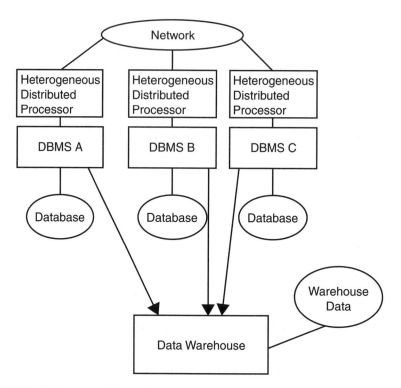

FIGURE B.19 Interoperability and warehousing.

B.9 FEDERATED DATABASES

As stated by Sheth and Larson,[37] a federated database system is a collection of cooperating but autonomous database systems belonging to a federation. The goal is for the database management systems that belong to a federation to cooperate with one another and yet maintain some degree of autonomy. Note that to be consistent with the terminology, we distinguish between a federated database management system and a federated database system. A federated database system includes both a federated database management system, the local DBMSs, and the databases. A federated database management system is that component which manages the different databases in a federated environment.

Figures B.20 and B.21 illustrate a federated database system. Database systems A and B belong to federation F1, while database systems B and C belong to federation F2. We can use the architecture illustrated in Figure B.18 for a federated database system. In addition to handling heterogeneity, the HDP also has to handle the federated environment (see Figure B.21). That is, techniques have to be adapted to handle cooperation and autonomy. We have called such an HDP a federated distributed processor (FDP). An architecture for an FDS is illustrated in Figure B.21.

Figure B.22 illustrates an example of an autonomous environment. There is communication between components A and B and between B and C. Due to autonomy,

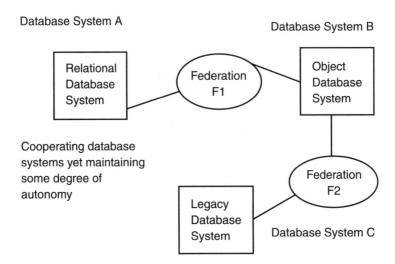

FIGURE B.20 Federated database management.

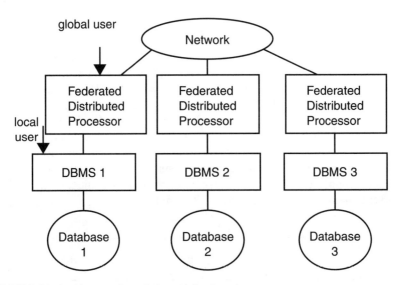

FIGURE B.21 Architecture for a federated database system.

it is assumed that components A and C do not wish to communicate with each other. Component A may get requests from its own user or from component B. In that case, it has to decide which request to honor first. Also, there is a possibility for component C to get information from component A through component B. In such a situation, component A may have to negotiate with component B before it gives a reply. The developments to deal with autonomy are still in the research stages. The challenge is to handle transactions in an autonomous environment. Transitioning the research into commercial products is also a challenge.

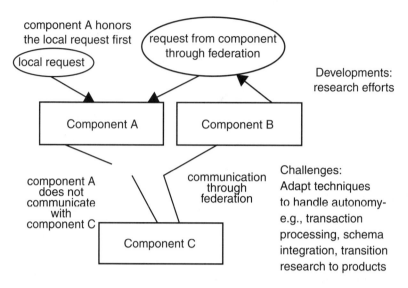

FIGURE B.22 Autonomy.

B.10 CLIENT–SERVER DATABASES

Earlier sections described interoperability between heterogeneous database systems and focused on the federated database systems approach. In that approach, different database systems cooperatively interoperate with each other. This section describes another aspect of interoperability based on the client–server paradigm. Major database system vendors have migrated to an architecture called the client–server architecture. With this approach, multiple clients access the various database servers through some network. A high level view of client–server communication is illustrated in Figure B.23. The ultimate goal is for multi-vendor clients to communicate with multi-vendor servers in a transparent manner. A specific example of client–server communication is illustrated in Figure B.24.

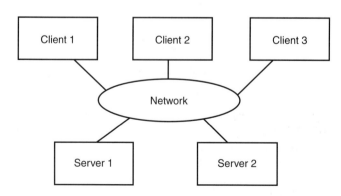

FIGURE B.23 Client–server architecture-based interoperability.

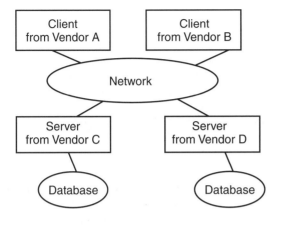

Developments:
research
prototypes,
commercial
products, ISO's
RDA standard,
OMG's CORBA,
SQL Access
Group's CLI

Challenges:
transaction
processing,
integrity and
security,
implementing
standards

FIGURE B.24 Example client–server architecture.

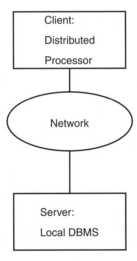

FIGURE B.25 An approach to place the modules.

One of the major challenges in client–server technology is to determine the modules of the distributed database system that need to be placed at the client and server sides. Figure B.25 shows an approach where all the modules of the distributed processor of Figure B.17 are placed at the client side, while the modules of the local DBMS are placed at the server side. Note that with this approach, the client does a lot of processing, which is called the fat client approach. There are also other options. For example, some of the modules of the distributed processor could be part of the server, in which case the client would be thinner.

In order to facilitate the communication between multiple clients and servers, various standards are being proposed. One example is the International Standards Organization's (ISO) remote database access (RDA) standard. This standard provides a generic interface for communication between a client and a server. Microsoft Corporation's open database connectivity (ODBC) is also becoming increasingly popular for clients to communicate with the servers. OMG's CORBA provides specifications for client–server communication based on object technology.[33] Here, one possibility is to encapsulate the database servers as objects, and the clients issue appropriate requests and access the servers through an object request broker. Other standards include IBM's DRDA (distributed relational database access) and the SQL Access Group's call level interface (CLI).* While many of the developments have been in query processing, the challenges are in transaction processing, semantic heterogeneity, integrity, and security.

In a previous book,[39] we described various aspects of client–server interoperability, in particular, technical issues for client–server interoperability, architectural approaches, three of the standards proposed for communication between clients and servers such as RDA, ODBC, and CORBA, as well as metadata aspects. Orfali et al.[34,35] provide a good reference. We will revisit distributed object management systems such as CORBA in the appendix on object management.

B.11 MIGRATING LEGACY DATABASES AND APPLICATIONS

Many database systems developed twenty to thirty years ago are becoming obsolete. Those systems use older hardware and software. Between now and the next few decades, many of today's information systems and applications will also become obsolete. Due to resource and, in certain cases, budgetary constraints, new developments of next generation systems may not be possible in many areas.[3] Therefore, current systems need to become easier, faster, less costly to upgrade, and less difficult to support. Legacy database system and application migration is a complex problem, and many of the efforts underway are still not mature. While a good book has been recently published on this subject,[8] there is no uniform approach for migration. Since migrating legacy databases and applications is becoming a necessity for most organizations, both government and commercial, we can expect a considerable amount of resources to be expended in this area in the near future. The research issues are also not well understood.

Migrating legacy applications and databases also has an impact on heterogeneous database integration. Typically, a heterogeneous database environment may include legacy databases as well as some of the next generation databases. In many cases, an organization may want to migrate the legacy database system to an architecture like the client–server architecture and still want the migrated system to be part of the heterogeneous environment. The functions of the heterogeneous database system may be impacted due to this migration process.

Two candidate approaches have been proposed for migrating legacy systems. One is to do all of the migration at once. The other is incremental migration, that

---

* Now part of the Open Group. Note that the products and standards have evolved over the years and are continually changing. We encourage the reader to keep up with the developments. Much of the information can be obtained on the Web.

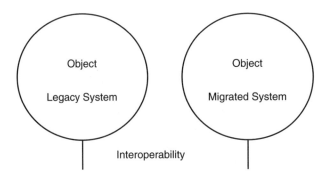

FIGURE B.26 Migrating legacy databases.

is, as the legacy system gets migrated, the new parts have to interoperate with the old parts. Various issues and challenges to migration are discussed by Thuraisingham.[39] Figure B.26 illustrates an incremental approach to migrating legacy databases through the use of object request brokers.

B.12 IMPACT OF THE WEB

The explosion of the users on the Internet and the increasing number of World Wide Web servers with large quantities of data are rapidly advancing database management on the Web. For example, heterogeneous information sources have to be integrated so that users access the servers in a transparent and timely manner. Security and privacy is becoming a major concern. So are other issues such as copyright protection and ownership of the data. Policies and procedures have to be set up to address these issues.

Database management functions for the Web include those such as query processing, metadata management, storage management, transaction management, security, and integrity. Thuraisingham[41] has examined various database management system functions and discussed the impact of Internet database access on these functions. Figure B.27 illustrates applications accessing various database systems on the Web.

B.13 SUMMARY

This appendix has discussed various aspects of database systems and provided a lot of background information to understand the various chapters in this book. We began with a discussion of various data models. We chose relational and entity-relationship models because they are more relevant to what we have addressed in this book. Then we provided an overview of various types of architectures for database systems, including functional and schema architectures. Next, we discussed database design aspects and administration issues. This appendix also provided an overview of the various functions of database systems, including query processing, transaction management,

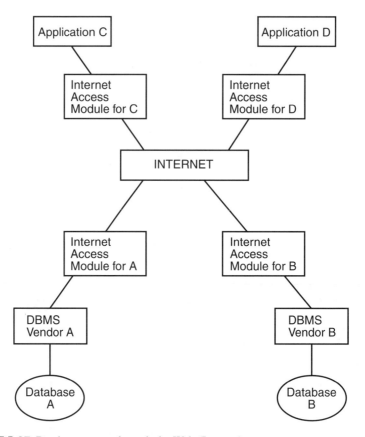

FIGURE B.27 Database access through the Web (Internet).

storage management, metadata management, security, integrity, and fault tolerance.*
Finally, we briefly discussed distributed databases and interoperability.

Many of the chapters in this book discuss various data management and data
mining system aspects related to multimedia information processing. These include
query processing, storage management, metadata management, security, distribution,
and interoperability.

REFERENCES

1. Special Issue on Federated Databases, *ACM Comput. Surv.,* September 1990.
2. Bell, D. and Grimson, J., *Distributed Database Systems,* Addison-Wesley, Reading,
   MA, 1992.
3. Bensley, E. et al., Evolvable systems initiative for real-time command and control
   systems, Proceedings of the First IEEE Complex Systems Conference, Orlando, FL,
   November 1995.

---

* Various texts and articles ave been published on database systems.[5-7,9,11,13,14,22,26,27,36,39-41,43]

4. Bernstein, P. et al., *Concurrency Control and Recovery in Database Systems,* Addison-Wesley, Reading, MA, 1987.

5. Brodie, M. et al., *On Conceptual Modeling: Perspectives from Artificial Intelligence, Databases, and Programming Languages,* Springer-Verlag, New York, NY, 1984.

6. Brodie, M. and Mylopoulos, J., *On Knowledge Base Management Systems,* Springer-Verlag, New York, NY, 1986.

7. Brodie, M. et al., *Readings in Artificial Intelligence and Databases,* Morgan Kaufmann, San Mateo, CA, 1988.

8. Brodie M. and Stonebraker, M., *Migrating Legacy Databases,* Morgan Kaufmann, San Mateo, CA, 1995.

9. Ceri, S. and Pelagatti, G., *Distributed Databases, Principles and Systems,* McGraw-Hill, New York, NY, 1984.

10. Chen, P., The entity relationship model — toward a unified view of data, *ACM Trans. Database Syst.,* March 1976.

11. Chorafas, D., *Intelligent Multimedia Databases,* Prentice-Hall, Englewood Cliffs, NJ, 1994.

12. Codd, E. F., A relational model of data for large shared data banks, *Commun. ACM,* May 1970.

13. Das, S., *Deductive Databases and Logic Programming,* Addison-Wesley, Reading, MA, 1992.

14. Date, C. J., *An Introduction to Database Management Systems,* Addison-Wesley, Reading, MA, 1990.

15. von Halle, B. and Kull, D., Eds., *Data Management Handbook,* Auerbach Publications, New York, NY, 1994.

16. von Halle, B. and Kull, D., Eds., *Data Management Handbook Supplement,* Auerbach Publications, New York, NY, 1995.

17. Thuraisingham, B., Ed., *Data Management Handbook Supplement,* Auerbach Publications, New York, NY, 1996.

18. Thuraisingham, B., Ed., *Data Management Handbook Supplement,* Auerbach Publications, New York, NY, 1998.

19. Proceedings of the 1994 DoD Database Colloquium, San Diego, CA, August 1994.

20. Proceedings of the 1994 DoD Database Colloquium, San Diego, CA, August 1995.

21. Elmasri, R., The entity category relationship model, *Data Knowledge Eng. J.,* March 1985.

22. Frost, R., *On Knowledge Base Management Systems,* Collins Publishers, London, U.K., 1986.

23. Special Issue on Heterogeneous Database Systems, *IEEE Comput.,* December 1991.

24. Proceedings of the IFIP Database Security Conferences, 1988–1994.

25. Kim, W. et al., *Query Processing in Database Systems,* Springer-Verlag, New York, NY, 1985.

26. Korth, H. and Silberschatz, A., *Database System Concepts,* McGraw-Hill, New York, NY, 1986.

27. Lloyd, J., *Logic Programming,* Springer-Verlag, Heidelberg, 1987.

28. Loomis, M., *Object Databases,* Addison-Wesley, Reading, MA, 1995.

29. Maier, D., *Theory of Relational Databases,* Computer Science Press, Rockville, MD, 1983.

30. Proceedings of the IEEE Metadata Conference, Silver Spring, MD, April 1996.

31. MIT Technical Reports on Data Quality.

32. *Object Database Standard: ODMB 93,* Object Database Management Group, Morgan Kaufmann, San Mateo, CA, 1993.

33. *Common Object Request Broker Architecture and Specification,* OMG Publications, John Wiley and Sons, New York, NY, 1995.

34. Orfali, R. et al., *Essential, Client Server Survival Guide,* John Wiley and Sons, New York, NY, 1994.

35. Orfali, R. et al., *The Essential, Distributed Objects Survival Guide,* John Wiley and Sons, New York, NY, 1994.

36. Prabhakaran, B., *Multimedia Database Systems,* Kluwer Publishers, Norwood, MA, 1997.

37. Sheth A. and Larson, J., Federated Database Systems, *ACM Comput. Surv.,* September 1990.

38. SQL3, American National Standards Institute, Draft, October 1992.

39. Thuraisingham, B., *Data Management Systems Evolution and Interoperation,* CRC Press, Boca Raton, FL, 1997.

40. Thuraisingham, B., *Data Mining: Technologies, Techniques, Tools and Trends,* CRC Press, Boca Raton, FL, 1998.

41. Thuraisingham, B., *Web Information Management and Electronic Commerce,* CRC Press, Boca Raton, FL, 2000.

42. Tsichritzis, D. and Lochovsky, F., *Data Models,* Prentice-Hall, Englewood Cliffs, NJ, 1982.

43. Ullman, J. D., *Principles of Database and Knowledge Base Management Systems,* Vols. I and II, Computer Science Press, Rockville, MD, 1988.

44. Wiederhold, G., Mediators in the architecture of future information systems, *IEEE Comput.,* March 1992.

45. Yang, D. and Torey, T., A practical approach to transforming extended ER diagrams into the relational model, *Inf. Sci. J.,* October 1988.

Appendix C
Data Mining

C.1 OVERVIEW

Data mining is another important technology for the Web. The increasing number of databases on the Web have to be mined to extract useful information. Data mining is the process of posing various queries and extracting useful information, patterns, and trends from large quantities of data possibly stored in databases. Essentially, for many organizations, the goals of data mining include improving marketing capabilities, detecting abnormal patterns, and predicting the future based on past experiences and current trends. There is clearly a need for this technology. There are large amounts of current and historical data being stored. Therefore, as databases become larger, it becomes increasingly difficult to support decision making. In addition, the data could be from multiple sources and domains. There is a clear need to analyze the data to support planning and other functions of an enterprise.

Various terms have been used to refer to data mining, as shown in Figure C.1. These include knowledge/data/information discovery and extraction. Note that some define data mining to be the process of extracting previously unknown information, while knowledge discovery is defined as the process of making sense of the extracted information. This book does not differentiate between data mining and knowledge discovery. It is difficult to determine whether a particular technique is a data mining technique. For example, some argue that statistical analysis techniques are data mining techniques. Others argue that they are not and that data mining techniques should uncover relationships that are not straightforward. For example, with data mining, a medical supply company could increase sales by targeting its advertising toward certain physicians who are likely to buy the products, or a credit bureau may limit its losses by selecting candidates who are not likely to default on their payments. Such real-world experiences have been reported in various papers.[8] In addition, data mining can also be used to detect abnormal behavior. For example, an intelligence agency can determine abnormal behavior of its employees using this technology.

Some of the data mining techniques include those based on rough sets, inductive logic programming, machine learning, and neural networks, among others. The data mining problems include classification (finding rules to partition data into groups), association (finding rules to make associations between data), and sequencing (finding rules to order data). Essentially, one arrives at some hypothesis, which is the information extracted, from examples and patterns observed. These patterns are observed by posing a series of queries; each query may depend on the responses obtained to the previous queries posed. There have been several developments in data mining. These include tools by corporations such as Lockheed Martin, Inc.[18]

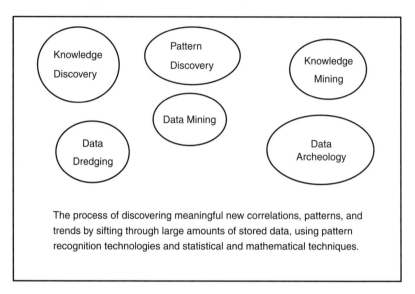

FIGURE C.1 Different definitions of data mining.

This chapter is organized as follows. Technologies that contribute to data mining are discussed in Section C.2. Essential concepts in data mining including techniques are discussed in Section C.3. Trends in data mining are the subject of Section C.4. Note that our previous book, *Data Mining: Technologies, Techniques,Tools and Trends*, is an elaboration of each of the three Sections: C.2, C.3, and C.4. Section C.5 describes data warehousing and its relationship to data mining. The impact of the Web is the subject of Section C.6. The chapter is summarized in Section C.7.

C.2 DATA MINING TECHNOLOGIES

Data mining is an integration of multiple technologies, as illustrated in Figure C.2. These include data management such as database management, data warehousing, statistics, machine learning, decision support, and other technologies such as visualization and parallel computing.* We briefly discuss the role of each of these technologies. It should be noted, however, that while many of these technologies, such as statistical packages and machine learning algorithms, have existed for many decades, the ability to manage and organize the data has played a major role in making data mining a reality.

Data mining research is being carried out in various disciplines. Database management researchers are taking advantage of the work being done on deductive and intelligent query processing for data mining. One of the areas of interest is extending query processing techniques to facilitate data mining. Data warehousing is also

---

* We have distinguished between data management and database management and also between data, information, and knowledge. Definitions of all of these terms are given in Appendix A as well as by Thuraisingham.[20]

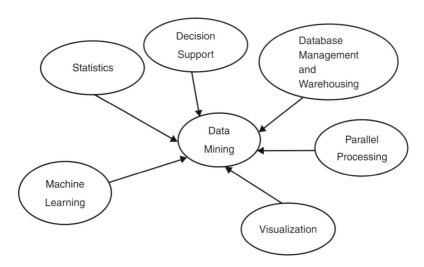

FIGURE C.2 Data mining technologies.

another key data management technology for integrating the various data sources and organizing the data so that it can be effectively mined.

Researchers in statistical analysis are integrating their techniques with machine learning techniques to develop more sophisticated statistical techniques for data mining. Various statistical analysis packages are now being marketed as data mining tools. There is some dispute over this. Nevertheless, statistics is a major area contributing to data mining.

Machine learning has been around for a while. The idea is for the machine to learn various rules from the patterns observed and then apply these rules to solve the problems. While the principles used in machine learning and data mining are similar, with data mining, one usually considers larger quantities of data. Therefore, integration of database management and machine learning techniques are needed for data mining.

Researchers from the computer visualization field are approaching data mining from another perspective. One of their areas of focus is using visualization techniques to aid the data mining process. In other words, interactive data mining is a goal of the visualization community.

Decision support systems are collections of tools and processes to help managers make decisions and guide them in management. For example, tools for scheduling meetings, organizing events, spreadsheets, graph tools, and performance evaluation tools are examples of decision support systems. Decision support has theoretical underpinnings in decision theory.

Finally, researchers in the high performance computing area are also working on developing appropriate techniques so that data mining algorithms are scalable. There is also interaction with the hardware researchers so that appropriate hardware can be developed for high performance data mining.

Several other technologies are beginning to have an impact on data mining, including collaboration, agents, and distributed object management. A discussion of all of these technologies is beyond the scope of this book. We have limited our focus to some of the key technologies. Furthermore, we emphasize that having good data is key to good mining.

C.3 CONCEPTS AND TECHNIQUES IN DATA MINING

There are a series of steps involved in data mining. These include getting the data organized, determining the desired outcomes, selecting tools, carrying out the mining, pruning the results so that only the useful ones are considered further, taking actions from the mining, and evaluating the actions to determine benefits. These steps are discussed in detail in this book. Some of the outcomes and techniques are briefly reviewed.

There are various types of data mining. By this we do not mean the actual techniques used to mine the data, but what the outcomes of the data mining will be. Some of these outcomes are discussed by Agrawal et al.[4] and are also addressed in this book. They have also been referred to as data mining tasks. We describe a few of them here.

In one outcome of data mining called classification, records are grouped into some meaningful subclasses. For example, suppose an automobile sales company has some information that all the people on its list who live in City X own cars worth more than $20,000. They may then assume that even those who are not on their list but live in City X can afford cars costing more than $20,000. This way, the company classifies the people living in City X.

A second outcome of data mining is sequence detection. By observing patterns in the data, sequences are determined. An example of sequence detection is, "after John goes to the bank, he generally goes to the grocery store."

A third outcome of data mining is data dependency analysis, in which potentially interesting dependencies, relationships, or associations between data items are detected. For example, if John, James, and William have a meeting, then Robert will also be at that meeting. This type of data mining is of much interest to many.

A fourth outcome of mining is deviation analysis. For example, "John went to the bank on Saturday, but he did not go to the grocery store after that. Instead, he went to a football game." With this type of outcome, anomalous instances and discrepancies are found.

As mentioned earlier, various techniques are used to obtain the outcomes of data mining. These techniques could be based on rough sets, fuzzy logic, inductive logic programming, or neural networks, among others, or they could simply be some statistical technique. We discuss these techniques elsewhere in this book. Furthermore, different approaches have also been proposed to carry out data mining, including top-down as well as bottom-up mining. Data mining outcomes, techniques, and approaches are illustrated in Figure C.3 and will be elaborated later in this book.

Numerous developments have been made in data mining over the past few years. Many of these focus on relational databases. That is, the data is stored in relational databases and mined to extract useful information and patterns. We have several research prototypes and commercial products. The research prototypes include those developed at IBM's Almaden Research Center and at Simon Fraser University. The

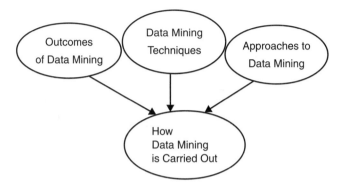

FIGURE C.3 Aspects of data mining.

prototypes and products employ various data mining techniques including neural networks, rule based reasoning, and statistical analysis. The various data mining tools in the form of prototypes and products will also be discussed in this book.

C.4 DIRECTIONS AND TRENDS IN DATA MINING

While several developments have been made, there are also many challenges. For example, due to the large volumes of data, how can the algorithms determine which technique to select and what type of data mining to do? Furthermore, the data may be incomplete or inaccurate. At times there may be redundant information, and at other times there may not be sufficient information. It is also desirable to have data mining tools that can switch to multiple techniques and support multiple outcomes. Some of the current trends in data mining include the following and are illustrated in Figure C.4:

- Mining distributed, heterogeneous, and legacy databases
- Mining multimedia data
- Mining data on the World Wide Web
- Security and privacy issues in data mining
- Metadata aspects of mining

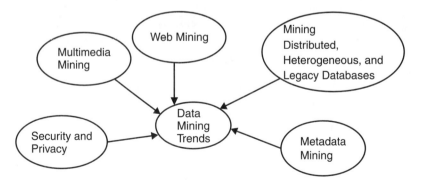

FIGURE C.4 Data mining trends.

In many cases, the databases are distributed and heterogeneous in nature. Furthermore, much of the data is in legacy databases. Mining techniques are needed to handle these distributed, heterogeneous, and legacy databases. Next, current data mining tools operate on structured data. However, there are still large quantities of data that are unstructured. Data in multimedia databases are often semistructured or unstructured. Data mining tools have to be developed for multimedia databases. The explosion of data and information on the World Wide Web necessitates the development of tools to manage and mine the data so that only useful information is extracted. Therefore, developing mining tools for the World Wide Web is an important area of work. Privacy issues are becoming critical for data mining.[19] Users now have sophisticated tools to make inferences and deduce information to which they are not authorized. Therefore, while data mining tools help solve many problems in the real world, they may also invade the privacy of individuals. Throughout our previous book,[20] we repeatedly stressed the importance of metadata for data management. Metadata also plays a key role in data mining.[21]

In addition to the trends in the above areas, there are also several challenges. These include handling dynamic, sparse, incomplete, and uncertain data, as well as determining which data mining algorithm to use and on what data to operate. In addition, mining multiple languages is also a challenge. Researchers are addressing these challenges.

C.5 DATA WAREHOUSING AND ITS RELATIONSHIP TO DATA MINING

Data warehousing is one of the key data management technologies to support data mining. Several organizations are building their own warehouses. Commercial database system vendors are marketing warehousing products. In addition, some companies are specializing only in developing data warehouses. What exactly is a data warehouse? The idea behind it is that it is often cumbersome to access data from heterogeneous databases. Several processing modules need to cooperate with each other to process a query in a heterogeneous environment. Therefore, a data warehouse will bring together the essential data from the heterogeneous databases. That way, the users need to query only the warehouse.

As stated by Inmon,[9] data warehouses are subject-oriented. Their design depends to a great extent on the application utilizing them. They integrate diverse and possibly heterogeneous data sources. They are persistent. That is, the warehouse is very much like a database. Data warehouses vary with time because as the data sources from which the warehouse is built get updated, the changes have to be reflected in the warehouse. Essentially, data warehouses provide support to decision support functions of an enterprise or an organization. For example, while the data sources may have the raw data, the data warehouse may have correlated data, summary reports, and aggregate functions applied to the raw data.

Figure C.5 illustrates a data warehouse. The data sources are managed by database systems A, B, and C. The information in these databases is merged and put into a warehouse. There are various ways to merge the information. One is to simply replicate the databases. This does not have any advantages over accessing the

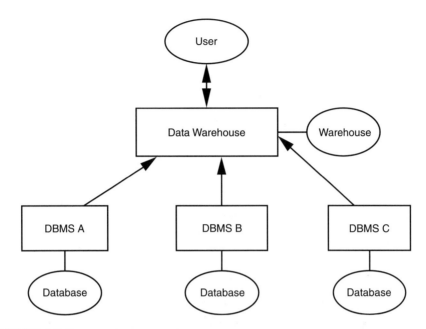

FIGURE C.5 Data warehouse example.

heterogeneous databases. The second option is to replicate the information but remove any inconsistencies and redundancies. This has some advantages, as it is important to provide a consistent picture of the databases. The third approach is to select a subset of the information from the databases and place it in the warehouse. There are several issues involved. How are the subsets selected? Are they selected at random or is some method used to select the data? For example, one could take every other row in a relation (assuming it is a relational database) and store these rows in the warehouse. The fourth approach, which is a slight variation of the third approach, is to determine the types of queries that users might pose, then analyze the data and store only the data that is required by the user. This is called on-line analytical processing (OLAP), as opposed to on-line transaction processing (OLTP).

With a data warehouse, data may often be viewed differently by different applications. That is, the data is multi-dimensional. For example, the payroll department may want data in a certain format, while the project department may want data in a different format. The warehouse must provide support for such multi-dimensional data.

In integrating the data sources to form the warehouse, one challenge is to analyze the application and select appropriate data to be placed in the warehouse. At times, some computations may have to be performed so that only summaries and averages are stored in the data warehouse. Note that the warehouse does not always have all the information for a query. In that case, the warehouse may have to get the data from the heterogeneous data sources to complete the execution of the query. What happens to the warehouse when the individual databases are updated? How are the updates propagated to the warehouse? How can security be maintained? These are some of the issues that are being investigated.

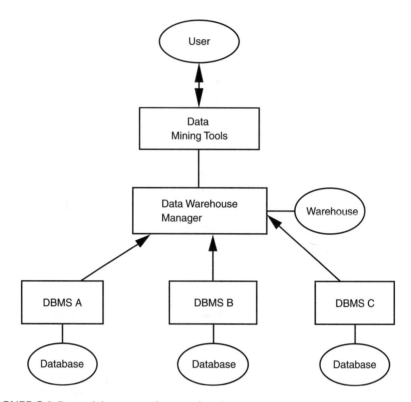

FIGURE C.6 Data mining versus data warehousing.

In our tutorials on data warehousing and mining, we are often asked which is preferable, building a warehouse or integrating the data sources? For example, what is the difference between warehousing and interoperability? A warehouse is built for decision support. Therefore, the data in the warehouse may not reflect the changes made in the data sources in a timely fashion. If the data sources are changing rapidly and if the user wants to see the changes, then a warehouse may not make much sense and one may want to simply integrate the heterogeneous data sources. However, in many cases, one could have both, that is, a warehouse as well as database systems that interoperate with each other.

We are also often asked the question as to the difference and/or relationship between warehousing and mining. Note that while data warehousing formats and organizes the data to support management functions, data mining attempts to extract useful information as well as predict trends from the data. Figure C.6 illustrates the relationship between data warehousing and data mining. Note that having a warehouse is not necessary to do mining because data mining can also be applied to databases. However, warehouses structure the data in such a way as to facilitate mining; so, in many cases it is highly desirable to have a data warehouse to carry out mining. The relationship between warehousing, mining, and database systems is illustrated in Figure C.7.

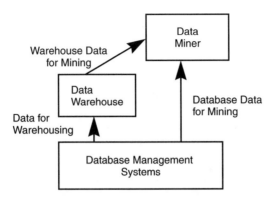

FIGURE C.7 Database systems, data warehousing, and mining.

Defining the point at which warehousing ends and mining begins is rather subjective. There are certain questions that warehouses can answer. Furthermore, warehouses have built-in decision support capabilities. Some warehouses carry out predictions and trends. In that case, warehouses carry out some of the data mining functions. In general, we believe that in the case of a warehouse, the answers to queries are found in the database. The warehouse has to come up with query optimization and access techniques to get the answer. For example, consider questions like, "how many red cars did physicians buy in 1990 in New York?" The answer is in the database. However, for a question like, "how many red cars do you think physicians will buy in 2005 in New York?," the answer may not be in the database. Based on the buying patterns of physicians in New York and their salary projections, one might be able to predict the answer to this question.

Essentially, a warehouse organizes data effectively so the data can be mined. The question is, is a data warehouse absolutely necessary to mine data? We believe that it is very good to have a warehouse, but that does not mean we must have a warehouse to mine. A good DBMS that manages a database effectively can also be used. Also, with a warehouse, one often does not have transactional data. Furthermore, the data may not be current; therefore, the results obtained from mining may not be current. If one needs up-to-date information, one can mine the database managed by a DBMS, which also has transaction processing features. Mining data that change often is a challenge. Typically, mining has been used for decision support data. Therefore, there are several issues that need further investigation before we can carry out real-time data mining. For now, at least, we believe that having a good data warehouse is critical to perform good mining for decision support functions. Note that one can also have an integrated tool that carries out both data warehousing and mining functions. We call such a tool a data warehouse miner, as illustrated in Figure C.8.

C.6 IMPACT OF THE WEB

Mining data on the Web is one of the major challenges facing the data management and mining community as well as those working on Web information management

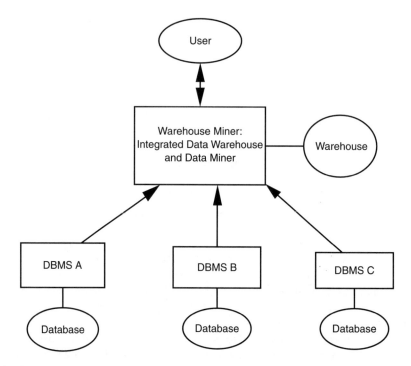

FIGURE C.8 Integrated data warehousing and data mining.

and machine learning. There is so much data and information on the Web that extracting the useful and relevant information for the user is the real challenge. Scanning the Web can become quite daunting, and one can easily get overloaded with data. The question is, how do we convert this data into information and, subsequently, knowledge so that the user gets only what he wants? Furthermore, what are the ways to extract information previously unknown from the data on the Web? More importantly how can mining the usage patterns on the Web improve the capability of an organization? In other words, electronic commerce is one of the major beneficiaries of Web mining. Figure C.9 illustrates how data mining tools may be applied to Web databases.

C.7 SUMMARY

This chapter has provided an introduction to data mining. We first discussed various technologies for data mining, and then we provided an overview of the concepts of data mining. These concepts include the outcomes of mining, the techniques employed, and the approaches used. The directions and trends, such as mining heterogeneous data sources, mining multimedia data, mining Web data, metadata aspects, and privacy issues, were addressed next. Finally, we provided an overview of data warehousing and its relationship to data mining.

Note that this book has given just enough information for the reader to understand some data mining topics related to the Web such as Web mining and Web privacy.

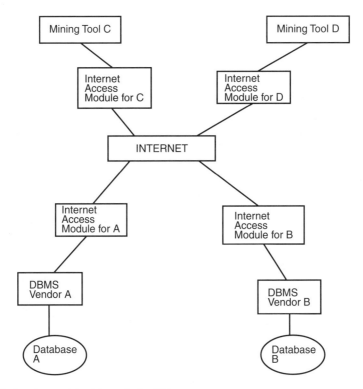

FIGURE C.9 Data mining through the Web.

For more details, refer to Thuraisingham.[1,22] Many important topics were covered by Thuraisingham,[21] so the reader has some idea of what data mining is all about. For an in-depth understanding of the various topics in data mining, we also refer the reader to the numerous papers and articles that have appeared in data mining and related areas.[1,7,10-16,23,24] We reference many of these throughout this book. In addition, data mining papers have also appeared at various data management conferences.[6,17,25] Recently, a federal data mining symposium series has been established.[3] In addition to Thuraisingham,[21] books on data mining include those by Adriaans and Zantinge[2] and Berry and Linoff.[5]

REFERENCES

1. Special Issue on Data Mining, *Commun. ACM,* November 1996.
2. Adriaans, P. and Zantinge, D., *Data Mining,* Addison-Wesley, Reading, MA, 1996.
3. Proceedings of the First Federal Data Mining Symposium, Washington, D.C., December 1997.
4. Agrawal, A. et al., Database mining a performance perspective, *IEEE Trans. Knowledge Data Eng.,* December 1993.
5. Berry, M. and Linoff, G., *Data Mining Techniques,* John Wiley and Sons, New York, NY, 1997.

6. Proceedings of the 1998 Data Engineering Conference, Orlando, FL, February 1998.

7. Fayyad, U. et al. *Advanced in Knowledge Discovery and Data Mining,* MIT Press, Cambridge, MA, 1996.

8. Grupe, F. and Owrang, M., Database mining tools, in *Handbook of Data Management Supplement,* Thuraisingham, B., Ed., Auerbach Publications, New York, NY, 1998.

9. Inmon, W., *Building the Data Warehouse,* John Wiley and Sons, New York, NY, 1993.

10. Proceedings of the First Knowledge Discovery in Databases Conference, Montreal, Canada, August 1995.

11. Proceedings of the Second Knowledge Discovery in Databases Conference, Portland, OR, August 1996.

12. Proceedings of the Third Knowledge Discovery in Databases Conference, Newport Beach, CA, August 1997.

13. Proceedings of the Fourth Knowledge Discovery in Databases Conference, New York, NY, August 1998.

14. Proceedings of the Knowledge Discovery in Databases Conference, Singapore, February 1997.

15. Proceedings of the Second Knowledge Discovery in Databases Conference, Melbourne, Australia, April 1998.

16. Proceedings of the ACM SIGMOD Workshop on Data Mining, Montreal, Canada, May 1996.

17. Proceedings of the 1998 ACM SIGMOD Conference, Seattle, WA, June 1998.

18. Simoudis, E. et al., Recon Data Mining System, Technical Report, Lockheed Martin Corporation, May 1995.

19. Thuraisingham, B., Data warehousing, data mining, and security, Proceedings of the Tenth IFIP Database Security Conference, Como, Italy, 1996.

20. Thuraisingham, B., *Data Management Systems Evolution and Interoperation,* CRC Press, Boca Raton, FL, 1997.

21. Thuraisingham, B., *Data Mining: Technologies, Techniques, Tools and Trends,* CRC Press, Boca Raton, FL, 1998.

22. Thuraisingham, B., *Web Data Management and Electronic Commerce,* CRC Press, Boca Raton, FL, 2000.

23. Special Issue on Data Mining, *IEEE Trans. Knowledge Data Eng.,* December 1993.

24. Special Issue on Data Mining, *IEEE Trans. Knowledge Data Eng.,* December 1996.

25. Proceedings of the Very Large Database Conference, New York, NY, August 1998.

Appendix D
Object Technology

D.1 OVERVIEW

Object technology, also referred to as OT or OOT (object-oriented technology), encompasses different technologies, including object-oriented programming languages, object database management systems, object-oriented design and analysis, distributed object management, components and frameworks. The underlying theme for all these object technologies is the object model. That is, the object model is the very essence of object technology. Any object system is based on some object model, whether it is a programming language or a database system. The interesting aspect of an object model is that everything in the real world can be modeled as an object.

The organization of this chapter is as follows. Section D.2 describes the essential properties of object models (OODM). Object-oriented programming languages (OOPL) are the subject of Section D.3. Object database systems will be discussed in Section D.4 (OODB). Object-oriented design and analysis (OODA) will be discussed in Section D.5. Distributed object management (DOM) is the subject of Section D.6. Section D.7. describes components and frameworks (C&F). Finally, Section D.8 discusses the impact of the Web. More specifically, the impact of object technology on Web data management will be discussed. The chapter is summarized in Section D.9. Figure D.1 illustrates the various types of object technologies. The OODM is common to all these technologies.

D.2 OBJECT DATA MODELS

Since the birth of object technology sometime during the 1970s, numerous object models have been proposed. In fact, some recent object models trace back to the language Simula in the 1960s. Initially, those models were to support programming languages such as Smalltalk. Later, those models were enhanced to support database systems as well as other complex systems. This section provides an overview of the essential feature of object models. Note that many of the features presented here are common for object models developed for different types of systems such as programming languages, database systems, modeling and analysis, and distributed object management systems.

While there are no standard object models, the unified modeling language (UML) proposed by the prominent object technologists (Rumbaugh, Booch, and Jacobson) has gained increasing popularity and has almost become standard in recent years. Our discussion of the object model has been influenced by much of our work on

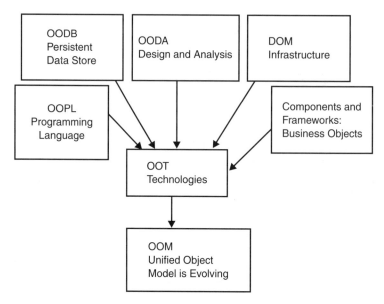

FIGURE D.1 Object technologies.

object database systems as well as the model proposed by Won Kim et al.[2] We call this model an object-oriented data model.*

The key points in an object-oriented model are encapsulation, inheritance, and polymorphism. With an object-oriented data model, the database is viewed as a collection of objects.[2] Each object has a unique identifier called the object ID (OID). Objects with similar properties are grouped into a class. For example, employee objects are grouped into EMP class, while department objects are grouped into DEPT class, as shown in Figure D.2. A class has instance variables describing the properties. The instance variables of EMP are SS#, Ename, Salary, and D#, while the instance variables of DEPT are D#, Dname, and Mgr. The objects in a class are its instances. As illustrated in Figure D.2, EMP has three instances and DEPT has two instances.

A key concept in object-oriented data modeling is encapsulation. That is, an object has well defined interfaces. The state of an object can only be accessed through interface procedures called methods. For example, EMP may have a method called Increase Salary. The code for Increase Salary is illustrated in Figure D.2. A message, for example, Increase Salary (1, 10K), may be sent to the object with the object ID of 1. The object's current salary is read and updated by 10,000.

A second key concept in an object model is inheritance, where a subclass inherits properties from its parent class. This feature is illustrated in Figure D.3, where the EMP class has MGR (manager) and ENG (engineer) as its subclasses. Other key

---

* Two types of object models have been proposed for databases. One is the object-oriented data model proposed for object-oriented databases and the other is the object-relational data model proposed for object-relational databases. We discuss the object-oriented data model in this section. Object-relational models are discussed in the section on object database management.

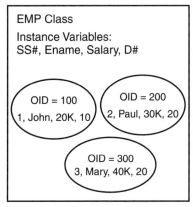
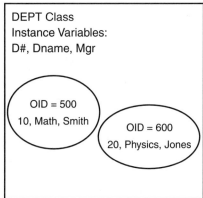

EMP Class

Instance Variables:
SS#, Ename, Salary, D#

OID = 100
1, John, 20K, 10

OID = 200
2, Paul, 30K, 20

OID = 300
3, Mary, 40K, 20

DEPT Class

Instance Variables:
D#, Dname, Mgr

OID = 500
10, Math, Smith

OID = 600
20, Physics, Jones

Increase-Salary (OID, Value)

Read-Salary(OID, Amount)

Amount : =Amount + Value

Write-Salary(OID, Amount)

FIGURE D.2 Objects and classes.

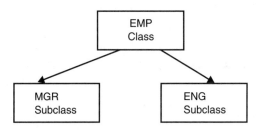

FIGURE D.3 Class–subclass hierarchy.

concepts in an object model include polymorphism and aggregation.* These features are discussed by Banerjee et al.[2] Further information can also be obtained in Thuraisingham.[10] Note that a second type of inheritance is when the instances of a class inherit the properties of that class.

A third concept is polymorphism, which is where one can pass different types of arguments for the same function. For example, to calculate area, one can pass a sphere or a cylinder object. Operators can be overloaded also. That is, the add operation can be used to add two integers or real numbers.

Another concept is the aggregate hierarchy, also called the composite object or the is-part-of hierarchy. In that case, an object has component objects, for example, a book object has component section objects and a section object has component paragraph objects. Aggregate hierarchy is illustrated in Figure D.4.

---

* Inheritance is also known as the is-a hierarchy. Aggregation is also known as the is-part-of hierarchy.

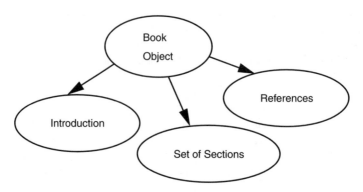

FIGURE D.4 Aggregate object.

Objects also have relationships between them. For example, an employee object has an association with the department object, which is the department in which he works. Also, the instance variables of an object could take integers, lists, arrays, or even other objects as values. All of these concepts are discussed in the book by Cattell.[3] The Object Data Management Group is also proposing standards for object data models.[6]

D.3 OBJECT-ORIENTED PROGRAMMING LANGUAGES

Object-oriented programming languages (OOPL) essentially go back to Simula in the 1960s. However, they really became popular with the advent of Smalltalk by the Xerox Palo Alto Research Center in the late 1970s. Smalltalk is a pure object-oriented programming language where everything is considered to be an object. Implementations of Smalltalk were being developed throughout the 1980s. Around the mid-1980s, languages such as LISP and C were made object-oriented by extending them to support objects. One such popular extension is the language C++. In the early to mid-1990s, a lot of programming was carried out in C++.

Around the 1990s, Sun Microsystems wanted to develop a language for its embedded computing and appliance business that would not have all of the problems associated with C++ such as pointers. The resulting language, first named Oak, was eventually called Java. Java became immensely popular because of the Internet. The language developers at Sun realized that Java, because of its write-once run-any-where property, could be an appropriate language for Internet programming. Today, we find that there is a huge demand for Java programmers, and the numbers of Java programmers are increasing rapidly.

Figure D.5 illustrates the evolution of object-oriented programming languages. Because of all of the features that object technology offers, for example, reuse because of inheritance, we believe that many of the future systems will be based on objects. For example, with inheritance, one can reuse code and modules for various systems. This is because software modules can inherit properties from other modules implemented as part of the class–subclass hierarchy.

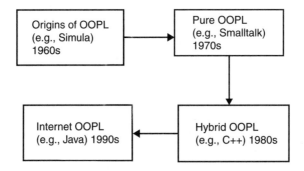

FIGURE D.5 Evolution of OOPL.

D.4 OBJECT DATABASE MANAGEMENT

D.4.1 OVERVIEW

This section discusses three types of object database systems. One is object-oriented database systems, which make object-oriented programming languages persistent. The second is extended-relational systems, which extend relational database systems with object layers. The third type is object-relational systems, where objects are nested within relations. Each of these are discussed in the following sections.

D.4.2 OBJECT-ORIENTED DATABASE SYSTEMS

Object-oriented database management systems (OODBMS) were developed to make programming languages persistent. The early systems such as Gemstone and Object-store made languages such as Smalltalk and C++ persistent. Gemstone was originally designed to make Smalltalk persistent, while Objectstore was designed to make C++ persistent. The idea was that tight integration with programming languages would be better for programming intensive applications, rather than having a loose coupling between application programs and SQL-based relational databases. Tight integration between OODBMSs and application programs is illustrated in Figure D.6. OODBMSs were also designed to support large and variable sized data blocks, multimedia data types, long-term lock and checkout of objects, high performance, and schema evolution.[1,5] These OODBMSs are currently in their second generation and can support relationships. The major focus at present is to make these OODBMSs the persistent store for Web data management as well as distributed object management systems.

D.4.3 EXTENDED-RELATIONAL SYSTEMS

While OODBMSs make programming languages persistent, relational systems are being extended to support objects. It is well understood that object management capability is needed to support complex data structures. At the same time, languages

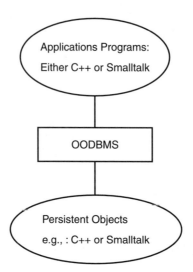

FIGURE D.6 OODBMS.

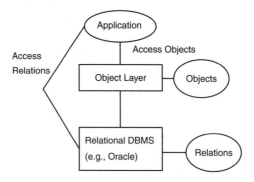

FIGURE D.7 Extended-relational database system.

such as SQL are standards languages. Therefore, one needs the capability to manipulate both relations and objects. One approach is to extend relational database
systems with object layers, as shown in Figure D.7. In this approach, applications
can access either objects or relations. Transformations between objects and relations
are performed by the object manager. That is, the object database is, in a way, a
virtual database. Objects eventually have to be transformed to relations, and eventually, the relations in the relational database are accessed through the relational
DBMS.

Various relational database vendors such as Oracle, Sybase, and Informix are
extending their database systems to support objects. These systems have been around
for quite a few years and can support complex data structures and multimedia data
types. While relational database vendors migrate toward objects, OODBMS vendors
support relationships and SQL-like query languages. We believe that eventually there
will be a middle ground between relations and objects.

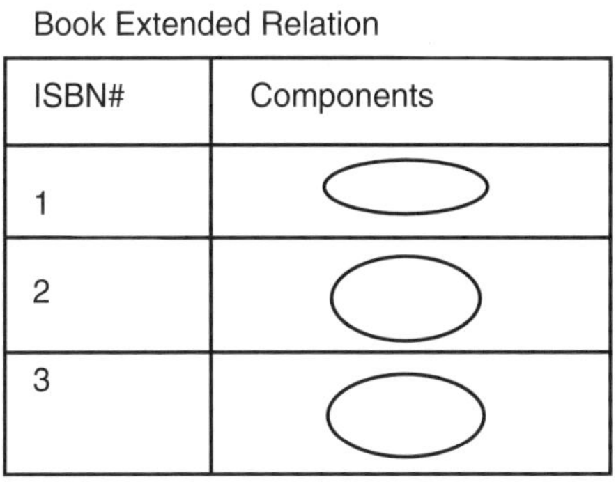

Book Extended Relation

ISBN#	Components
1	⬭
2	◯
3	◯

FIGURE D.8 Object-relational model.

D.4.4 OBJECT-RELATIONAL SYSTEMS

Object-relational database systems were developed to overcome some of the problems with relational and object-oriented database systems. The relational data model is based on well-defined principles. Furthermore, a notable feature of relational database systems is the query language. SQL, developed initially for relational databases, is an ANSI standard. However, relational data models cannot support complex objects, which are needed for new generation applications such as CAD/CAM and multimedia. On the other hand, object-oriented data models can support complex structures. However, in general, object-oriented database systems do not provide good support for querying.

To overcome these problems, relational database vendors are building some sort of support for objects, as discussed in the earlier section. Object-oriented database vendors are developing better query interfaces as well as better support to represent relationships. In addition, a third kind of system, an object-relational database system, has been developed. These systems provide support both for relations and objects. Note that there is no standard object-relational model. With one approach, the relations are extended so that the data elements are no longer atomic. That is, the data elements could be complex objects. Figure D.8 illustrates this concept, i.e., where the book relation has an attribute called components. This attribute describes the components of the book.

Object-relational database systems are essentially database systems that manage object relations. These database systems support SQL extended to support complex constructs as well as object languages that have SQL-like capabilities. Object-relational systems are still young, and we can expect them to mature over the next few years. Several object-oriented database system products, as well as a few object-relational database system prototypes, are discussed in the literature.[1]

D.5 OBJECT-ORIENTED DESIGN AND ANALYSIS

The previous two sections addressed object-oriented programming languages and object-oriented database management. In the 1980s, there was a lot of interest in

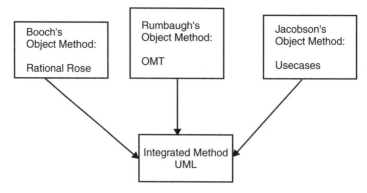

FIGURE D.9 OODA approaches.

using object technology to design and analyze applications. Prior to that, various analysis techniques such as structured analysis and Jackson diagrams were used to analyze the application. At the same time, entity relationship models were very popular to represent the entities of the application and the relationships between them.

With the advent of OOT, interest in using objects to model and analyze applications increased. Various design and analysis methodologies were proposed. Notable among them were the method suggested by Booch, Usecases by Jacobson, and OMT (object modeling technique) by Rumbaugh et al. There was so much debate about these methods that at the 1993 OOPSLA (object-oriented programming systems, languages, and applications) Conference there was an extremely contentious discussion about this subject.

Surprisingly, within the following two years, it was announced that the three groups (Booch, Jacobson, and Rumbaugh et al.) were merging and producing a unified methodology called UML (unified modeling language).[4] UML has essential features from the three original approaches and is now more or less a standard for object modeling and analysis. Figure D.9 illustrates the convergence of the three approaches. Essentially, in object-oriented design and analysis methodologies like, for example, OMT, an object model, similar to the one we proposed in this appendix, is used to represent the static parts of the application. A dynamic model is used to capture the interactions between the entities and the timing and synchronization of events, and a functional model generates the methods. In addition, there is a system design phase in which the various modules of the system are identified, and, during the object design phase the algorithms are designed. The idea of the usecase method is to capture the user requirements with objects and do rapid prototyping. By combining all of the good features of the three prominent methods, UML has become the choice modeling methodology.

A closely related development in the mid-1990s was the design patterns methodology. The design patterns methodology is all about extracting patterns in the modeling methodology that are often reused for various applications. For example, consider a general ledger application which would use objects and communication between objects in a standard way. Once this pattern is captured, it can be reused

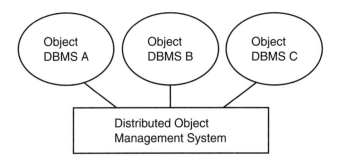

FIGURE D.10 Interoperability based on distributed object management.

for various general ledger applications by different users and corporations. The design patterns methodology has simply exploded as a concept, and various people have published their patterns for different systems and applications. There are also Web-based user and discussion groups on patterns.

D.6 DISTRIBUTED OBJECT MANAGEMENT

D.6.1 OVERVIEW

Various types of distributed object management systems for interoperability have been proposed in the literature. We discussed them briefly in a previous work.[10] Section D.6.1 provides an overview of distributed objects, and Section D.6.2 discusses a special distributed object management approach, the Object Management Group's CORBA (common object request broker architecture).

D.6.2 DISTRIBUTED OBJECT MANAGEMENT APPROACH

Distributed object management (DOM) technology is being used increasingly to interconnect heterogeneous databases, systems, and applications. With this approach, the various systems and applications are encapsulated as objects, and the objects communicate with each other through exchanging messages. Figure D.10 provides a high level view of interoperability based on DOM technology. In Figure D.10, components A, B, and C are encapsulated as objects and communicate with each other through the DOM system.*

DOM technology can also be used for finer-grained encapsulation. For example, mediators, repositories, and DBMSs can be encapsulated so that different DBMSs interoperate. This is illustrated in Figure D.11. Key here is that each system must have well defined interfaces using a common interface definition language. What is inside each encapsulated object is transparent to the remote object. This technology is also being used for migrating legacy databases and applications. For a detailed discussion of DOM technology, refer to Orfali et al.[9] Examples of DOM technology include OMG's CORBA and Microsoft's DCOM, and, more recently, Sun Microsystems' JINI can also be regarded as a DOM technology. Since this text is not

---

* Internet-based component integration is discussed in Part II of this book.

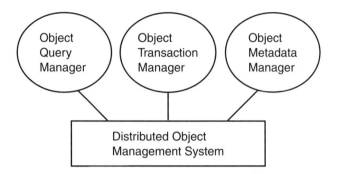

FIGURE D.11 Distributed object management for component integration.

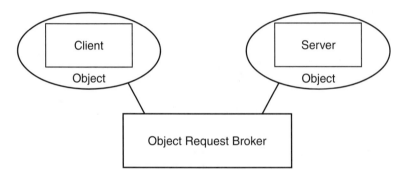

FIGURE D.12 Interoperability through ORB.

about DOM, in the next section discusses only CORBA. Again, we have selected CORBA because we are most familiar with this approach.

D.6.3 CORBA

An example of a distributed object management system used as a middleware to connect heterogeneous database systems is a system based on OMG's CORBA.* CORBA is a specification that enables heterogeneous applications, systems, and databases to interoperate with each other. There are three major components to CORBA.[7] One is the object model which essentially includes most of the constructs discussed in Chapter 2, the second is the object request broker (ORB) through which clients and servers communicate with each other, and the third is the interface definition language (IDL) which specifies the interfaces for client–server communication. Figure D.12 illustrates client–server communication through an ORB. In that figure, the clients and servers are encapsulated as objects. The two objects then communicate with each other through the ORB.

Furthermore, the interfaces must conform to IDL. Since heterogeneous database management is of interest to us, our goal is for heterogeneous database systems to

---

* Note that middleware is referred to as the intermediate layer which lies between the operating systems and the applications. This layer connects different systems and applications.

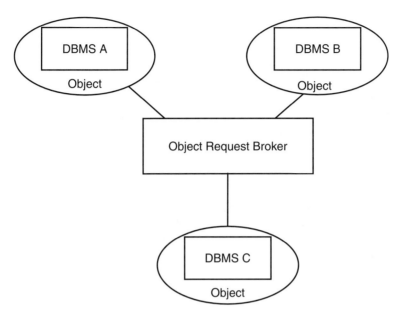

FIGURE D.13 Heterogeneous database integration through the ORB.

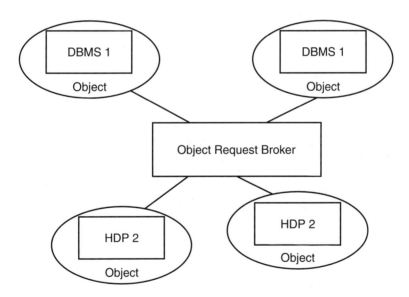

FIGURE D.14 Finer-grained encapsulation.

interoperate with each other through the ORB. Figure D.13 illustrates this integration. Essentially, nodes A, B, and C are encapsulated as objects that communicate
through the ORB. Note that this example assumes a coarse-grained encapsulation
where entire nodes are encapsulated as objects. One could, however, encapsulate
portions of the modules. For example, Figure D.14 illustrates a case where the

DBMSs and HDPs (i.e., the heterogeneous distributed processor discussed by Thuraisingham[10]) are encapsulated separately. In Figure D.14, we have four objects, two for DBMSs and two for HDPs. One can continue this way and obtain even finer-grained encapsulation where modules of the DBMS and modules of the HDP are encapsulated. The advantage of finer-grained encapsulation is that it facilitates migration. That is, one can throw away an HDP and replace it with newer modules. However, the more objects there are, the more messages are sent through the ORB, which will have an impact on the performance.

Various special interest groups and task forces are coming up with specifications based on CORBA. These include specifications for security, real-time processing, and Internet access. In addition, there is also work being carried out on developing specifications for vertical domains such as medical, financial, and transportation domains. OMG is also developing specifications for various business objects. It appears that this technology shows a lot of promise for interoperability. A 1994 workshop[8] addressed the state of CORBA technology and whether it was ready for prime time.* However, since then, CORBA technology has matured, and several products now exist. Furthermore, OMG is actively involved in specifying services for areas such as security, real-time, and fault tolerance.

D.7 COMPONENTS AND FRAMEWORKS

Components and frameworks make up one of the latest object technologies and have really taken off since the mid-1990s. The terms components and frameworks have no standard definitions. There was an excellent survey of the field in the October 1997 issue of *Communications of the ACM*.[11] In a sense, a framework can be considered a skeleton with classes and interconnections. One then instantiates this skeleton for various applications. Frameworks are being developed for different application domains including financial and medical.

Components, on the other hand, are classes, objects, and relationships between them that can be reused. Components can be built for different applications, for example, there are components for financial, medical, and telecommunication applications. These components are also called business objects. One can also develop general purpose components. For example, a database system can be developed as a collection of components for query processing, transaction management, and metadata management. Figure D.15 illustrates components for data management. That figure assumes that various vendors produce different components for data management that have to be integrated to produce a DBMS. Figure D.16 illustrates components for three-tier client–server processing.

---

* Note that OMG is a consortium of over 800 corporations. Other DOM technologies that are becoming increasingly popular are Microsoft Corporation's Distributed OLE (object linking and embedding) and COM (component object model). The future of these technologies will be clearer in the next few years.[9] Thuraisingham[10] has provided a brief overview of Microsoft's database access product OLE/DB. Note that the names of the various distributed object management system products are also evolving.

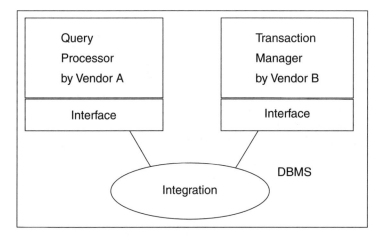

FIGURE D.15 Component integration for database management.

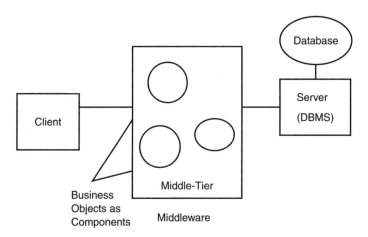

FIGURE D.16 Three-tier components.

D.8 IMPACT OF THE WEB

Now that we have explained all the key points of object technology, let us examine the impact of the Web. More specifically, let us examine how the Web is affected by object technology. First of all, object databases (OODB) can be used like any other database to interface to the Web. Object databases can also be a persistent store for various resources on the Web. Note that one can also use relational database systems. However, object database systems, with their representational power, are capable of storing richer data structures.

One area that has made a big impact is OOPL, and this great influence is because of Java. Java has basically become the standard programming language for the Web.

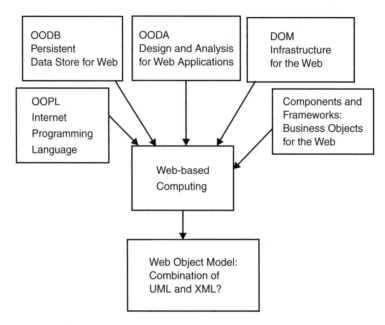

FIGURE D.17 Web-based computing and OT.

Various applications programs, applets, and servlets are being written in Java.* Java applets can be embedded in Web pages and executed when a Web page is brought into the browser environment from the server. This is a significant development. Object-oriented design and analysis methodologies are being used to design Web-based applications. Distributed object systems such as CORBA are being used as infrastructures for Web-based integration. For example, the Netscape browser uses ORBs for its infrastructure. Components and frameworks are being used for Web-based three-tier computing. One popular component technology is Sun Microsystem's Enterprise Java Beans (EJB). We now hear of various types of beans being developed for different Web-based applications, and these beans are based on EJB. Figure D.17 illustrates how objects are being impacted by the Web and vice versa.

D.9 SUMMARY

This appendix summarized some of the important object technology developments that have taken place in the past three decades. We started with object models, the very essence of object technology. We then discussed the evolution of OOPLs. This was followed by a discussion of various types of object database systems. Object-oriented design and analysis was given some consideration. Next, we provided an overview of distributed object management. Finally, the emerging area of components and frameworks was discussed. The impact of this technology on the Web is tremendous. This book demonstrates how object technology is being applied.

---

* Applets and servlets are explained later.

REFERENCES

1. Special Issue on Next Generation Database Systems, *Commun. ACM,* October 1991.
2. Banerjee, J. et al., A data model for object-oriented applications, *ACM Trans. Office Inf. Syst.,* October 1987.
3. Cattel, R., *Object Data Management Systems,* Addison-Wesley, Reading, MA, 1991.
4. Fowler, M. et al., *UML Distilled: Applying the Standard Object Modeling Language,* Addison-Wesley, Reading, MA, 1997.
5. Loomis, M., *Object Databases,* Addison-Wesley, Reading, MA, 1993.
6. *Object Database Standard: ODMB 93,* Object Database Management Group, Morgan Kaufmann, CA, 1993.
7. *Common Object Request Broker Architecture and Specification,* OMG Publications, John Wiley and Sons, New York, NY, 1995.
8. OOPSLA 94 Workshop on CORBA, Portland, OR, 1994.
9. Orfali, R. et al., *The Essential, Distributed Objects Survival Guide,* John Wiley and Sons, New York, NY, 1996.
10. Thuraisingham, B., *Data Management Systems Evolution and Interoperation,* CRC Press, Boca Raton, FL, 1997.
11. *Comun. ACM,* October 1997.

Appendix E
Data and Information Security

E.1 OVERVIEW

The number of computerized databases has increased rapidly during the past three decades. The advent of the Internet as well as networking capabilities has made access to data and information much easier. For example, users can now access large quantities of information in a short space of time. As more and more tools and technologies are being developed to access and use data, there is an urgent need to protect the data. Many government and industrial organizations have sensitive and classified data that has to be protected. Various other organizations such as academic institutions also have sensitive data about their students and employees. As a result, techniques for protecting the data stored in database management systems (DBMSs) have become an urgent need.

In the past three decades, various developments have been made to secure databases. Much of the early work was performed on statistical database security. Then, in the 1970s, as research in relational databases began, attention was directed toward access control issues. In particular, work on discretionary access control models began. While some work on mandatory security started in the late 1970s, it was not until the Air Force summer study in 1982 that many of the efforts in multilevel secure database management systems were initiated.[1] This resulted in the development of various secure database system prototypes and products. In the 1990s, with the advent of new technologies such as digital libraries, the World Wide Web, and collaborative computing systems, there was much interest in security not only in government organizations, but also in the commercial industry.

This chapter provides an overview of the various developments in information security with special emphasis on database security. Section E.2 discusses basic concepts such as access control for information systems. Section E.3 provides an overview of secure systems. Secure database systems are discussed in Section E.4. Since much of this book focuses on Web data management, we give some consideration to secure database systems. Emerging trends are the subject of section E.5. The impact of the Web is addressed in Section E.6. The chapter is summarized in Section E.7. For a detailed discussion of the information in this chapter, refer to Ferarri and Thuraisingham.[4]

E.2 ACCESS CONTROL AND OTHER SECURITY CONCEPTS

Access control models include those for discretionary security and mandatory security. In this section, we discuss both aspects of access control and also consider other issues.

In discretionary access control models, users or groups of users are granted access to data objects. These data objects could be files, relations, objects, or even data items. Access control policies include rules such as, "User U has read access to Relation R1 and write access to Relation R2." Access control could also include negative control such as, "User U does not have read access to Relation R."

In mandatory access control, subjects that act on behalf of users are granted access to objects based on some policy. A well known policy is the Bell and LaPadula policy[2] where subjects are granted clearance levels and objects have sensitivity levels. The set of security levels form a partially ordered lattice where Unclassified < Confidential < Secret < TopSecret. The policy has two properties, which are the following. A subject has read access to an object if its clearance level dominates that of the object. A subject has write access to an object if its level is dominated by that of the object.

Other types of access control include role-based access control, where access is granted to users depending on their roles and the functions they perform. For example, personnel managers have access to salary data, while project mangers have access to project data. The idea is to permit access on a need-to-know basis.

While the early access control policies were formulated for operating systems, those policies have been extended to include other systems such as database systems, networks, and distributed systems. For example, a policy for networks includes policies not only for reading and writing but also for sending and receiving messages.

Other security policies include administration policies. These policies include those for ownership of data as well as how to manage and distribute the data. Database administrators and system security officers are involved in formulating administration policies. Security policies also include policies for identification and authentication. Each user or subject acting on behalf of a user has to be identified and authenticated, possibly using some password mechanisms. Identification and authentication becomes more complex for distributed systems. For example, how can a user be authenticated at a global level?

The steps to developing secure systems include developing a security policy, developing a model of the system, designing the system, and verifying and validating the system. The methods used for verification depend on the level of assurance that is expected. Testing and risk analysis is also part of the process. These activities will eliminate or assess the risks involved. Figure E.1 illustrates various types of security policies.

E.3 SECURE SYSTEMS

The previous section discussed various policies for building secure systems. This section elaborates on various types of secure systems. Much of the early research in the 1960s and 1970s focused on securing operating systems. Early security policies, such as the Bell and LaPadula policy, were formulated for operating systems. Subsequently, secure operating systems such as Honeywell's SCOMP and MULTICS were developed.[5] Other policies, such as those based on noninterference, also emerged in the early 1980s.

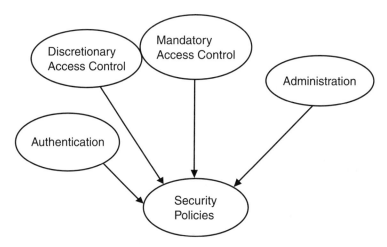

FIGURE E.1 Secure policies.

While early research on secure database systems was reported in the 1970s, it was not until the early 1980s that active research began in this area. Much of the focus was on multilevel secure database systems. The security policy for operating systems was modified slightly. For example, the write policy for secure database systems was modified to state that a subject has write access to an object if the subject's level is equal to that of the object. Since database systems enforced relationships between data and had semantics, there were additional security concerns. For example, data could be classified based on content, context, and time. The problem of posing multiple queries and inferring sensitive information from the legitimate responses became a concern. This problem is now known as the inference problem. Also, research was carried out not only on securing relational systems but on object systems as well.

Research on computer networks began in the late 1970s and continued throughout the 1980s and beyond. The networking protocols were extended to incorporate security features, and the result was secure network protocols. The policies include those for reading, writing, sending, and receiving messages. Research on encryption and cryptography has received much prominence due to networks and the Internet. Security for stand-alone systems was extended to include distributed sytems. These systems included distributed databases and distributed operating systems. Much of the research on distributed systems now focuses on securing the Internet, known as Web security, as well as securing systems such as distributed object management systems.

As new systems emerge, such as data warehouses, collaborative computing systems, multimedia systems, and agent systems, security for such systems has to be investigated. With the advent of the Internet and the World Wide Web, security is being given serious consideration by government organizations as well as commercial organizations. With e-commerce, it is important to protect the company's intellectual property. Figure E.2 illustrates various types of secure systems.

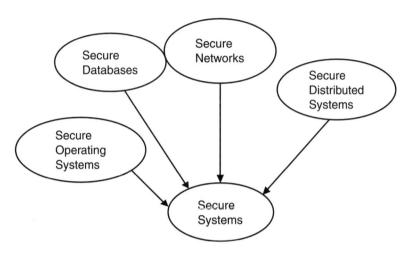

FIGURE E.2 Secure systems.

E.4 SECURE DATABASE SYSTEMS

Work on discretionary security for databases began in the 1970s when security aspects were investigated for System R at IBM's Almaden Research Center. Essentially, the security properties specified the read and write access that a user may have to relations, attributes, and data elements. In the 1980s and 1990s, security issues were investigated for object systems. The security properties specified the access that users had to objects, instance variables, and classes. In addition to read and write, method execution access was also specified.

Since the early 1980s, much of the focus of research was on multilevel secure database management systems. These systems essentially enforce the mandatory policy discussed in Section E.2 with the modification described in Section E.3. Since the 1980s, various designs, prototypes, and commercial products of multilevel database systems have been developed. Ferrari and Thuraisingham give a detailed survey of some of the developments.[4] Example efforts include SeaView by SRI International and Lock Data Views by Honeywell. These efforts extended relational models with security properties. One challenge was to design a model where a user sees different values at different security levels. For example, at the unclassified level, an employee's salary may be $20,000 and at the secret level it may be $50,000. In the standard relational model, such ambiguous values cannot be represented due to integrity properties.

Note that several other significant developments have been made regarding multilevel security for other types of database systems. These include security for object database systems.[9] In this effort, security properties specify read, write, and method execution policies. Much work was also carried out on secure concurrency control and recovery. The idea is to enforce security properties and still meet consistency without covert channels. Research was also carried out on multilevel security for distributed, heterogeneous, and federated database systems. Another area

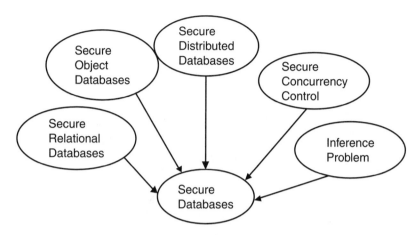

FIGURE E.3 Secure database systems.

that received a lot of attention was the inference problem. For details on the inference problem, refer to Thuraisingham.[11] For more information on secure concurrency control, refer to the numerous algorithms by Bertino and Jajodia.[3] For information on secure distributed and heterogeneous databases as well as secure federated databases, refer to Thuraisingham.[10,12]

As database systems become more sophisticated, securing these systems will become more and more difficult. Some of the current work focuses on securing data warehouses, multimedia databases, and Web databases.[6-8] Figure E.3 illustrates various types of secure database systems.

E.5 EMERGING TRENDS

In the mid-1990s, research on secure systems expanded to include emerging systems. These included securing collaborative computing systems, multimedia computing, and data warehouses. Data mining resulted in new security concerns. Since users now have access to various data-mining tools, the inference problem could be exacerbated because the data mining tool may make correlations and associations which may be sensitive. On the other hand, data mining could also help with security problems such as intrusion detection and auditing.

The advent of the Web resulted in extensive investigations of security for digital libraries and electronic commerce. In addition to developing sophisticated encryption techniques, security research also focused on securing Web clients as well as servers. Programming languages such as Java were designed with security in mind. Much research was also carried out on securing agents.

Secure distributed system research focused on security for distributed object management systems. Organizations such as OMG started working groups to investigate security properties. As a result, there are now secure distributed object management systems commercially available. Figure E.4 illustrates the various emerging secure systems and concepts. More details are given by Ferarri and Thuraisingham.[4]

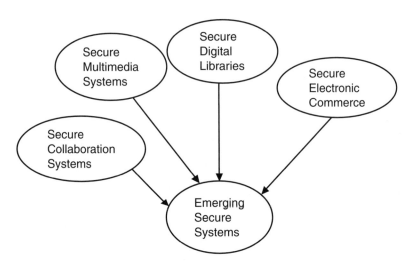

FIGURE E.4 Emerging secure systems.

E.6 IMPACT OF THE WEB

The advent of the Web has greatly impacted security. Security is now part of mainstream computing. Government organizations as well as commercial organizations are concerned about security. For example, in a financial transaction, millions of dollars could be lost if security is not maintained. With the Web, all sorts of information is available about individuals, and, therefore, privacy may be compromised.

Various security solutions are being proposed to secure the Web. In addition to encryption, the focus is on securing clients as well as servers. That is, end-to-end security has to be maintained. Web security also has an impact on electronic commerce. When one carries out transactions on the Web, it is critical that security is maintained. Information such as credit card numbers and social security numbers has to be protected.

All of the security issues discussed in the previous sections have to be considered for the Web. For example, appropriate security policies have to be formulated. This is a huge challenge since no one person owns the Web. The various secure systems, including secure operating systems, secure database systems, secure networks, and secure distributed systems, may be integrated in a Web environment. Therefore, this integrated system has to be secure. Problems like the inference problem may be exacerbated due to the various data mining tools. The various agents on the Web have to be secure. In certain cases, trade-offs need to be made between security and other features. That is, quality of service is an important consideration. In addition to technological solutions, legal aspects also have to be examined. That is, lawyers and engineers have to work together. Research on securing the Web is just beginning. Figure E.5 illustrates aspects of Web security.

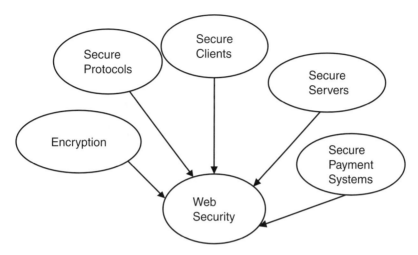

FIGURE E.5 Aspects of Web security.

E.7 SUMMARY

This appendix has provided a brief overview of the developments in secure systems. We discussed basic concepts in access control as well as discretionary and mandatory policies. We also provided an overview of secure systems. In particular, secure operating systems, secure databases, secure networks, and secure distributed systems were discussed. Additionally, we provided some details on secure databases. Finally, we discussed some research trends and the impact of the Web.

Future directions for secure database systems will be driven by the developments of the World Wide Web. Database systems are no longer stand-alone systems; they are being integrated into various applications such as multimedia, electronic commerce, mobile computing systems, digital libraries, and collaboration systems. Therefore, security issues for all these new generation systems will be very important. Furthermore, there are many developments for various object technologies such as distributed object systems and components and frameworks. Security for such systems is being investigated. Eventually, the security policies of the various subsystems and components have to be integrated to form policies for entire systems. There will be many challenges in formulating policies for such systems. New technologies such as data mining will help solve security problems such as intrusion detection and auditing. However, these technologies can also violate the privacy of individuals because adversaries can now use mining tools to extract unauthorized information about various individuals. Finally, migrating legacy databases and applications will continually be a challenge. Security issues for such operations cannot be overlooked.

Essentially, this chapter is a summary of a more detailed paper on secure database systems by Ferrari and Thuraisingham.[4]

REFERENCES

1. Air Force Summer Study Board Report on Multilevel Secure Database Systems, Department of Defense Document, Washington, D.C., August 1983.
2. Bell, D. and LaPadula, L., Secure Computer Systems: Unified Exposition and Multics Interpretation, MITRE Technical Report, Bedford, MA, 1975.
3. Bertino, E. and Jajodia, S., Secure concurrency control, *IEEE Trans. Knowledge Data Eng.*, August 1997.
4. Ferarri, E. and Thuraisinham, B., Database security: survey, in *Advanced Database Technology and Design*, Piattini, M. and Diaz, O., Eds., Artech House, Norwood, MA, 2000.
5. Special Issue on Computer Security, *IEEE Comput.*, 1983.
6. Proceedings of the IFIP Conference Series in Database Security, North Holland, 1988–1994.
7. Panel on data warehousing and data mining security, Proceedings of the IFIP 1997 Conference in Database Security, Thallasolicki, Greece, July 1998.
8. Panel on data mining and Web security, Proceedings of the IFIP 1998 Conference Series in Database Security, Thallasolicki, Greece, July 1998.
9. Thuraisingham, B., Security for object-oriented database systems, Proceedings of the ACM OOPSLA Conference, New Orleans, LA, October 1989.
10. Thuraisingham, B., Security for distributed database systems, *Comput. Security,* December 1991.
11. Thuraisingham, B. et al., Design and implementation of a database inference controller, *Data Knowledge Eng. J.,* December 1993.
12. Thuraisingham, B., Security for federated database systems, *Comput Security,* December 1994.

Appendix F
The World Wide Web, E-Business, and E-Commerce

F.1 OVERVIEW

The developments of the Internet have been key to the development of the World Wide Web. The Internet began as a research project funded by the United States Department of Defense. Much of the work was carried out in the 1970s. At that time, there were numerous developments with networking. Various networking protocols and products began to emerge. In addition, standards groups such as the International Standards Organzation proposed a layered stack of protocols for networking. The Internet research resulted in TCP/IP (transmission control protocol/internet protocol) for transport communication.

In the 1970s, while networking concepts were advancing rapidly, data management technology was emerging. Then, in the late 1980s, the early ideas of President Bush to organize and structure information started getting computerized. These ideas led to the development of hypermedia technologies. In the 1980s, researchers thought that these hypermedia technologies would result in efficient access to large quantities of information, for example, in library information systems. It was not until the early 1990s that researchers at CERN in Switzerland combined Internet and hypermedia technologies, which resulted in the World Wide Web. The idea behind the Web is for the various Web servers scattered within and across corporations to be connected through intranets and the Internet so that people from all over the world can have access to the right information at the right time. The advancement of various data and information management technologies contributed to the rapid growth of the World Wide Web.

One of the leading applications for the Web is e-commerce. This appendix discusses the evolution of the Web and then provides an introduction to e-commerce. For more details on this topic, refer to one of our previous books, *Web Data Management and Electronic Commerce*.[5] We address e-commerce as multimedia technology plays a key role in e-commerce and the broader area of e-business. Section F.2 provides an overview of the Web, while Section F.3 addresses e-commerce. The appendix is summarized in section F.4.

F.2 EVOLUTION OF THE WEB

The inception of the World Wide Web took place at CERN in Switzerland. Although different people have been credited with creating the Web, one of the early conceivers of the Web was Timothy Bernes-Lee who was working at CERN at that time. He now heads the World Wide Web Consortium (W3C) which specifies standards for the Web including data models, query languages, and security.

As soon as the Web emerged in the early 1990s, a group of graduate students at the University of Illinois developed a browser called MOSAIC. A company called Netscape Communications then marketed MOSAIC, and, since then, various browsers as well as search engines have emerged. These search engines, browsers, and servers all constitute today's World Wide Web. The Internet became the transport medium for communication.

Various protocols for communication such as HTTP (hypertext transfer protocol) and languages for creating Web pages such as HTML (hypertext markup language) also emerged. Perhaps one of the most significant developments is the Java programming language developed by Sun Microsystems. The work is now being continued by Javasoft, a subsidiary of Sun. Java is a language very much like C++ but avoids all the disadvantages of C++ like pointers. Java was developed as a programming language to be run platform-independent. It was soon found that this was an ideal language for the Web. Therefore, there are now various Java applications as well as what is known as Java applets. Applets are Java programs residing in a machine that can be called by a Web page running on a separate machine. Therefore, applets can be embedded into Web pages to perform all kinds of features. Of course, this involves additional security restrictions, because applets could come from untrustworthy machines. Another concept is a servlet. Servlets run on Web servers and perform specific functions such as delivering Web pages for a user request. Applets and servlets will be elaborated upon later in this appendix.

Middleware for the Web is continuing to evolve. If the entire environment is Java, connecting Java clients to Java servers, one could use RMI (remote method invocation) by Javasoft. If the platform consists of heterogeneous clients and servers, one could use Object Management Group's CORBA for interoperability. Some argue that client–server technology will die off because of the Web. That is, one may need different computing paradigms such as the federated computing model for the Web. These issues as well as various architectural issues are addressed in Part II of this book.

Another development for the Web is the area of components and frameworks. We discussed some of them in Chapter 5. Component technology, such as Enterprise Java Beans (EJB), is becoming very popular for componentizing various Web applications. These applications are managed by what is now known as application servers. These application servers (such as BEA's Web Logic) communicate with database management systems through data servers (these data servers may be developed by database vendors such as Object Design Inc.). Finally, one of the latest technologies for integrating various applications and systems, possibly heterogeneous, through the Web is Sun's JINI.[1] It essentially encompasses Java and RMI as its basic elements. Some of these technologies are addressed in Part II of this book.

The Web is continuing to expand and explode. Now there is so much data, information, and knowledge on the Web that managing all of it is becoming critical. Web information management is all about developing technologies for managing this information. One particular type of information system is a database system.

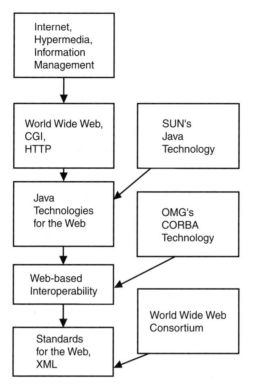

FIGURE F.1 World Wide Web.

Part II of this book provides some details on Web database management and discusses technologies for Web information management. Part III provides an overview of electronic commerce, the new way of doing business on the Web, and discusses the applications of Web information management to electronic commerce. Figure F.1 illustrates some of the Web concepts discussed here.

One of the major problems with the Internet is information overload. Because people can now access large amounts of information very rapidly, they can quickly become overloaded with information, and, in some cases, the information may not be useful to them. Furthermore, in certain other cases, the information may even be harmful to the users. The current search engines, although improving steadily, still give the users too much information. When a user types in an index word, many irrelevant Web pages are also retrieved. To solve this problem, we need intelligent search engines. The technologies that we have discussed in this book, if implemented successfully, could prevent this information overload problem. For example, agents may filter out information so that users get only the relevant information. Data mining technology could extract meaningful information from the data sources. Security technology could prevent users from getting information that they are not authorized to know. In addition to computer scientists, researchers in psychology, sociology, and other disciplines are also involved in examining various aspects of Internet database management. We need people in multiple disciplines to collaboratively work together

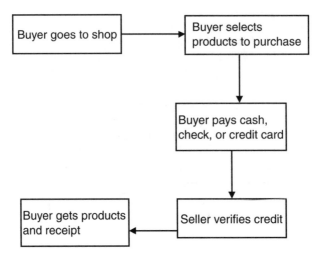

FIGURE F.2a Process of commerce.

to make the Internet a useful tool for its users. One of the emerging goals of Web technology is to provide appropriate support for data dissemination. This deals with getting the right data/information at the right time to the analyst/user (directly to the desktop if possible) to assist in carrying out various functions. Many of these technologies such as agents and data mining for the Web are addressed in Part II.

F.3   INTRODUCTION TO E-COMMERCE

F.3.1   OVERVIEW

This section discusses what is now referred to as the leading application for the Web, that is, e-commerce. What is e-commerce? Simply stated, it means carrying out commerce on the Web. Essentially, e-commerce involves transactions on the Web, i.e., buying and selling products on the Web. Earlier in this book, we mentioned that e-commerce could be as simple as putting up a Web page or as complicated as merging two corporations on the Web. More recently, we have heard the term e-business, a much broader term than e-commerce, which means doing any business on the Web. Therefore, e-commerce has come to be known as carrying out transactions on the Web, and tasks like putting up Web pages and other activities are part of e-business. Figure F.2a illustrates a normal business transaction (that is, a non-Web transaction), and Figure F.2b illustrates a business transaction on the Web.

This section provides a broad overview of e-commerce. We first discuss e-business and its relationship to e-commerce. This will give the reader some knowledge of the latest buzzwords in e-commerce. Then we discuss some models for e-commerce, in particular, business-to-business e-commerce as well as business-to-consumer e-commerce models.* It should be noted that models for e-commerce are rather immature,

---

* Business-to-business e-commerce is popularly called B-to-B, and business-to-consumer e-commerce is popularly called B-to-C.

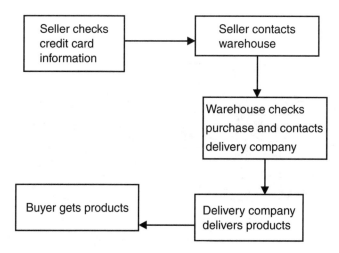

FIGURE F.2b Process of e-commerce.

and as we learn more about e-commerce, more highly evolved models will emerge. We also briefly discuss information technologies for e-commerce.

The organization of this chapter is as follows. Section F.3.2 discusses e-business and its relationship to e-commerce. Models of e-commerce are the subject of Section F.3.3. Information technologies for e-commerce are the subject of Section F.3.4.

F.3.2   E-BUSINESS AND E-COMMERCE

The term e-business is often heard today. Many companies prefer to be transacting e-business rather than e-commerce because they feel e-commerce may be too narrow, while e-business encompasses e-commerce. We believe that e-commerce can be considered broad such as putting up a Web page, listening to music on the Web, or conducting transactions on the Web. However, to be consistent with the emerging terminology, we will explain the differences between e-business and e-commerce. Note that the two terms are often used interchangeably.

Those who differentiate between e-business and e-commerce state that e-commerce involves only carrying out transactions on the Web. E-business, however, is much broader and includes learning and training on the Web, entertainment on the Web, putting up Web pages and hosting Web sites, conducting procurement on the Web, carrying out supply chain management, handling help on the Web for telephone repairs or other services, and almost anything that can be conducted on the Web. E-business and some of its various components are illustrated in Figure F.3.

Various types of corporations are now involved in e-business. One group consists of corporations that simply have Web pages. A second group consists of corporations that carry out e-commerce. A third group consists of corporations that help other corporations formulate e-business strategies. A fourth group consists of corporations that provide solutions and products for e-business. Other groups include those that carry out E-learning and E-training, E-procurement, and provide E-helpdesks. With

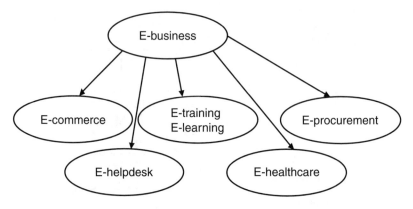

FIGURE F.3 E-business and its components.

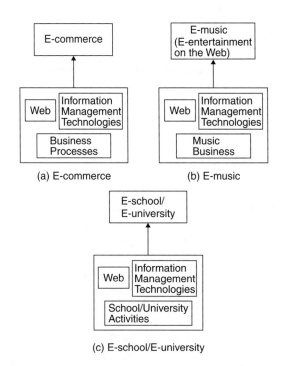

FIGURE F.4 Building blocks for e-business.

E-helpdesks, the time it takes to handle a customer's problem is greatly reduced and the need for too many human operators is eliminated.[2] Corporations that provide consulting as well as solutions and products include Fortune 100 corporations like IBM or smaller corporations like the dot-com companies. Some of these smaller corporations can connect consumers with healthcare providers, lawyers, real estate agents, and others who provide services. Consulting companies come in and assess the state of a corporation and its business practices and advise the corporation on how to develop e-business solutions. One of the latest trends is to provide fully integrated enterprise resource management and business process reengineering on the Web. Corporations such as SAP-AG are active in this area.

Figure F.4 illustrates the building blocks of e-business. For example, in Figure F.4a, the building blocks are the Web, information management technologies, and business processes (such as the business processes supported by the SAP product). These building blocks support e-commerce. Figure F.4b illustrates the building blocks for E-music (i.e., entertainment on the Web). These include the Web, information management technologies, and the music business. Figure F.4c illustrates building blocks for universities and schools. These include the Web, information management, and school/university business activities. To carry out good e-commerce, not only do we need the technologies described in this book, but we also need good business practices. We have approached the subject from a technology point of view since we are technologists and not business specialists. Nevertheless, business specialists are necessary to build an e-commerce organization.

F.3.3 MODELS FOR E-COMMERCE

As mentioned earlier, there are no well defined models for e-commerce. However, two paradigms, which we can consider models, are emerging. They are business-to-business e-commerce and business-to-consumer e-commerce. This section discusses both these models with examples.

As its name implies, business-to-business e-commerce is all about two businesses conducting transactions on the Web. Suppose corporation A is an automobile manufacturer and needs microprocessors to be installed in its automobiles. It will then purchase the microprocessors from corporation B who manufactures the microprocessors. Another example is when an individual purchases some goods such as toys from a toy manufacturer. This manufacturer then contacts a packaging company via the Web to deliver the toys to the consumer. The transaction between the manufacturer and the packaging company is a business-to-business transaction. Business-to-business e-commerce also involves one business purchasing a unit of another business or two businesses merging. The main point is that such transactions have to be carried out on the Web.

Business-to-consumer e-commerce is when a consumer makes purchases on the Web. In the toy manufacturer example above, the individual's purchase from the toy manufacturer is a business-to-consumer transaction. Business-to-consumer e-commerce has grown tremendously during the past year.[3,4] While computer hardware purchases still lead e-commerce transactions, purchasing toys, apparel, software, and even groceries via the Web have also increased. Many feel that the real future of e-commerce will be in business-to-business transactions because they involve millions of dollars.

The major difference between the two models is how business is carried out. This is similar to regular business transactions in the real word. In a live business-to-consumer transaction, people can give credit cards, cash, or checks to make a purchase. On the Web, credit cards are used most often. However, the use of E-cash and checks is also being investigated. In normal business-to-business transactions,

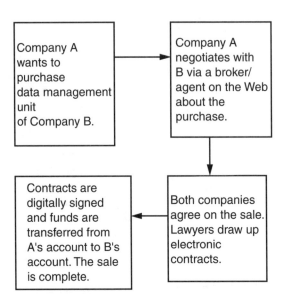

FIGURE F.5 Business-to-business e-commerce.

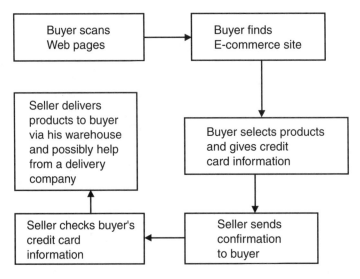

FIGURE F.6 Business-to-consumer e-commerce.

corporations have company accounts that are maintained and the corporations are billed at certain times. This is the approach being taken in the e-commerce world, too. That is, corporations have accounts with one another and these accounts are billed when purchases are made. Figures F.5 and F.6 illustrate examples of business-to-business and business-to-consumer transactions, respectively.

Regardless of the type of model, one of the major goals of e-commerce is to complete transactions on time. For example, in the case of business-to-consumer e-commerce, the seller has to minimize the time between the time of purchase and the time the buyer gets his goods. The seller may have to depend on third parties such as packaging and trucking companies to achieve this goal. It should also be noted that with e-commerce, the consumer has numerous choices for products. In a typical shop, the consumer does not have access to all of the products that are available. He cannot see the products displayed at the shop. However, in the world of e-commerce, the consumer has access to all the products that are available to the seller.

Another key point is the issue of trust. How can the consumer trust the seller and how can the seller trust the consumer? For example, the consumer may give his credit card number to a seller who is a fraud. The consumer may also be a fraud and not send a check when he gets the goods. The best known model is business/consumer relationships in the non-Web world. This is not always the case, however, in e-commerce. Some of the challenges involved in e-commerce are not very different from the mail order and catalog world. If the goods do not arrive, the consumer can write to his credit card company. But this may be a lengthy and possibly legal process. Another solution is for the seller to set up an account with a credit card company to establish its credibility. That is, a vendor from some unknown company may not be able to establish a relationship with a credit card company, and, therefore, the buyer may not be in danger of purchasing from a fraudulent source. In the e-commerce world, there are several additional security measures such as secure wallets and cards; these aspects were discussed by Thuraisingham.[5]

F.3.4 INFORMATION TECHNOLOGIES FOR E-COMMERCE

Without the various data and information management technologies, e-commerce cannot be a reality. That is, the technologies discussed in the various parts of this book are essentially technologies for e-commerce. E-commerce also includes nontechnological aspects such as policies, laws, social, and psychological impacts. We are now doing business in an entirely different way and, therefore, we need a paradigm shift. We cannot successfully transact E-commerce if we hold on the traditional method of buying and selling products. We have to be more efficient and rely on the technologies to gain a competitive edge.

Figure F.7 illustrates the overall picture of the technologies that may be applied to e-commerce. These include database systems, data mining, security, multimedia, interoperability, collaboration, knowledge management, and visualization. Details of how these technologies help e-commerce are given by Thuraisingham.[5]

F.4 SUMMARY

This appendix provided a broad overview of the Web and e-commerce. We started with a discussion of the evolution of the Web and then discussed the e-commerce process, followed by a discussion of the differences between e-business and e-commerce. Then we described models for e-commerce as well as information technologies for e-commerce. It should be noted that multimedia data management is a key technology for e-commerce.

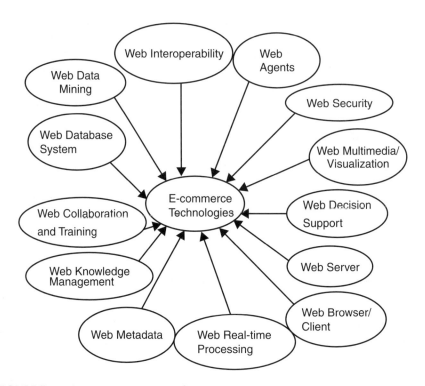

FIGURE F.7 Information technologies for e-commerce.

REFERENCES

1. *Commun. ACM,* May 1999.
2. *Bus. Week,* Asian Edition, December 1999.
3. *Inf. Week,* November 1999.
4. *Inf. World,* December 1999.
5. Thuraisingham, B., *Web Data Management and Electronic Commerce,* CRC Press, Boca Raton, FL, 2000.

Index